The Walking Whales

University of California Press, one of the most
distinguished university presses in the United States,
enriches lives around the world by advancing scholarship
in the humanities, social sciences, and natural sciences. Its
activities are supported by the UC Press Foundation and
by philanthropic contributions from individuals and
institutions. For more information, visit www.ucpress.edu.

University of California Press
Oakland, California

Library of Congress Cataloging-in-Publication Data

Thewissen, J. G. M., author.
 The walking whales : from land to water in eight
million years / J.G.M. Thewissen ; with illustrations by
Jacqueline Dillard.
   pages cm
 Includes bibliographical references and index.
 ISBN 978-0-520-27706-9 (cloth : alk. paper)—
 ISBN 978-0-520-95941-5 (e-book)
  1. Whales, Fossil—Pakistan. 2. Whales, Fossil—India.
3. Whales—Evolution. 4. Paleontology—Pakistan.
5. Paleontology—India. I. Title.
 QE882.C5T484 2015
 569'.5—dc23

                                    2014003531

Printed in China
23 22 21 20 19 18 17 16 15 14
10 9 8 7 6 5 4 3 2 1

The paper used in this publication meets the minimum
requirements of ANSI/NISO Z39.48–1992 (R 2002)
(*Permanence of Paper*).

Cover illustration (clockwise from top right):
Basilosaurus, Ambulocetus, Indohyus, Pakicetus, and
Kutchicetus. These are the animals that show that whales
once had land-living ancestors. The background of this
painting is a composite: these animals did not live in the
same habitat or the at the same time.

The Walking Whales

From Land to Water in Eight Million Years

J.G.M. "Hans" Thewissen

with illustrations by Jacqueline Dillard

UNIVERSITY OF CALIFORNIA PRESS

This book is dedicated to all the students, postdocs, fossil preparators, and technicians who worked in my lab and made this journey scientifically exciting as well as fun, in chronological order: Sandy, Ellen, Tony, Mary, Lauren, Mary Elizabeth, Amy, Lisa, Brooke, Sirpa, Rick, Bobbi Jo, Meghan, Sharon, Jenny, Denise, Summer, and Ashley. And it is dedicated to Elizabeth, for her encouragement. And it is dedicated to my mother, who supported everything I did enthusiastically.

Contents

*These six headings summarize the biology of the six fossil groups that form the transition between whales and their terrestrial ancestors. Their relationships to each other and to the living families of cetaceans (whales, dolphins, and porpoises) are given in figure 66.

CHAPTER I

A Wasted Dig

FOSSILS AND WAR

Punjab, Pakistan, January 1991. I am excited beyond belief! The National Geographic Society is giving me money to collect fossils in Pakistan: my very own field project, the first time ever. For years, it has been great to collect fossils in exotic places—Wyoming, Sardinia, and Colombia. But this is different. Now I can run my own program, decide where to collect, and study what is found. It's exciting but also daunting. My friend Andres Aslan will come with me. We're perfect complements: he loves geology and I love fossils. We're both just out of school, freshly minted PhDs, and together we're ready and able to set the world on fire, or at least vacuum up any fossil between Attock and Islamabad.

It is Andres's first trip to Pakistan. I first went there as a paleontology student in 1985, during the Soviet occupation of Afghanistan, when the CIA channeled much of its support against the Soviets through Pakistan. Trucks full of equipment would travel the highway, the Grand Trunk Road, from Islamabad to the Afghan border at night—the very same road we took to our field area. The Soviet-backed Afghan government retaliated by trying to disturb the stability of Pakistan. Car bombs were the weapon of choice, and my hotel room in Islamabad offered an excellent view into the courtyard of the police station next door, where a line of charred, exploded mini-busses were evidence of their success. With mirrors tied to long poles, the police stopped every vehicle entering the city and checked the underside for bombs. This didn't bother me,

FIGURE 1. Map of northern Pakistan and India, with places mentioned in this book. Fossil localities are indicated by bones (see also figure 22).

as long as I could collect fossils, studying life from a very exciting period in earth's history, fifty million years ago.

Now, six years later, Andres and I arrive in Pakistan just before the New Year and receive our permits to work in the Kala Chitta Hills, west toward the Afghan border (figure 1). The television in our Islamabad hotel is showing CNN stories about the Iraqi invasion of Kuwait last year, but that conflict seems distant. I am here to immerse us in the greatest excitement paleontology has to offer: collecting fossils, being the first to see and figure out each one I pick up.

We check in at a hotel in the town of Attock, and fieldwork starts on January 1. We travel to remote sites that I chose from decades-old reports from other paleontologists. The rocks here in the shadows of the Himalayas have their own distinctive charms. They are gnarled, bent, twisted, and overturned, all the result of the mountain-building to the north. They are silent witnesses to the incredibly violent forces that raised the Himalayas to be the world's highest mountains. With a sense of poetry, my Pakistani colleague, Mr. Arif, calls the limestone that has been tossed into tight bends "the dancing limestone."

We search the dry scrublands every day, but fossils are rare; things take time. In my fieldbook I have logged fifty-one fossils. None seem exciting—small pieces of fishbone, crocodile armor, fish teeth, and a piece of the casing of the ear of a whale, the tympanic bone. It is not the first whale bone I ever found. Growing up, in Holland, I lived close to a fossil locality where my father used to take me. A river had dropped rocks there that it collected as it cut through mountains upstream, in Belgium, France, and beyond. There was everything from sea lilies hundreds of millions of years old, to plant fossils from coal swamps, and large fossil whale bones from when that area was covered by ocean, just a few million years ago. It cemented my interest in fossils, and for my twelfth birthday I got a rock hammer, which is still the hammer I use now.

I have never studied whales before, and now too, whale bones are no good for me. The money from the National Geographic Society is for studying how land mammals migrated between Indo-Pakistan and Asia across the Tethys Sea some fifty million years ago. Whales are of no use for studying migrations on land. I need land-dwelling mammals, and many more fossils, if this grant is to be successful. I am very aware that failing to deliver on a first grant can sink a career.

On day five the dream collapses. The United States is threatening to invade Kuwait, and the U.S. government is worried about the safety of its citizens. Mr. Arif is told by his superiors at the Geological Survey of Pakistan to escort us back to Islamabad, the capital. All my plans are crumbling before my eyes. The reason for going back to Islamabad seems ridiculous—the conflict is in the Gulf, not Pakistan. The physical dangers seem much smaller than when I first visited. Why should politics end the field season?

Reluctantly, Andres and I return to Islamabad and check into a hotel. The hotel is in the Blue Area, Islamabad's broad central avenue with shopping areas, as well as the buildings of the president, prime minister, and congress. It's a mile or so from the American embassy.

We hang around our hotel room, waiting for news. On TV, the foreign minister of Iraq, Tariq Aziz, and the U.S. secretary of state, James Baker, are sparring. Mr. Arif tells us that we will be kicked out of Pakistan if war breaks out, and we will not be allowed to go back to the field. We visit the American consulate, pleading, hoping they will support our cause. The consulate is a fortress, with a concrete moat around it, double gates with Pakistani guards, and a second gate with U.S. marines and watch towers.

"It is laughable how often I have been attacked and misrepresented about this bear."[7] Although he was convinced that whales, being mammals, had ancestors that were derived from land mammals, the fossil record did not preserve any intermediates. All known fossil whales were obligate marine mammals. In Darwin's time, the oldest cetaceans known were basilosaurids—large whales with a streamlined shape easily recognizable to anyone familiar with modern whales. One hundred and thirty years later, when we found the *Pakicetus* incus, those were still the oldest whales for which skeletons were known.

But when those first basilosaurid skeletons were found, they were not immediately identified as whales. In 1832—before Darwin—twenty-eight giant vertebrae washed out of the banks of the Ouachita River in Louisiana. One of the vertebrae ended up with Dr. Richard Harlan in Philadelphia, who published an account of the find in 1834.[8] Harlan said that the vertebra pertained to a giant lizard. He called it *Basilosaurus*, after the Greek *basileus*, king, and *saurus*, lizard. This was a mistake—making an ancient aquatic mammal sound like a terrestrial lizard—but it was an understandable one. Whale vertebrae look different from those of land mammals, and Harlan had only one. Additional remains of a similar beast were found in 1834 and 1835 on an Alabama plantation. Harlan took these teeth and bones to London, showing them to the famous British zoologist Professor Richard Owen, a contemporary of Darwin and a critic of his evolutionary ideas. Owen recognized the teeth as clearly mammalian and noticed resemblances in the vertebrae of Harlan's beast to those of whales.[9] Feeling that Harlan's name was inappropriate, Owen renamed the animal *Zeuglodon cetoides*, after the yoke-like appearance of the teeth in side view (*zeugleh* means yoke in Greek, and the Latin *dens* means tooth; figure 6) and the whale-like appearance of its vertebrae (*cetoides*, whale-like).

The renaming was unfortunate. If Owen had lived today, he would not have given the fossil a new name, despite the problems with the original moniker. Biologists have realized that scientific names are vehicles for storing and retrieving information about animals and plants, and that the most important thing about them is that they are stable: everyone uses just one name for one animal. It does not matter whether the name describes the animal well. This is similar to human last names—after all, a person named Farmer may not be a farmer at all. The International Commission on Zoological Nomenclature has now established clear rules[10] specifying that, if two scientists have given different names to the same beast, the oldest name is valid. Today all species are named

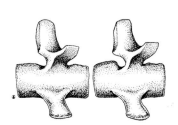

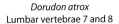

Dorudon atrox
Lumbar vertebrae 7 and 8

Basilosaurus cetoides
Lumbar vertebrae 7 and 8

Dorudon atrox
Lower second molar

FIGURE 6. Some fossils of extinct basilosaurid whales. Two *Basilosaurus* lumbar vertebrae (middle) show the large conical area called the centrum, but also a small protrusion, the neural arch (which sticks out on top). In mammals, the shape of the joints on the neural arch restricts the mobility between vertebrae. Given that these joints are small in *Basilosaurus*, it must have had a very flexible back. Compare these vertebrae to those of its close relative *Dorudon* (left), in which the neural arches lock the vertebrae in place, leading to a less flexible spine. A lower molar (right) shows the yoke-like appearance of the teeth of *Basilosaurus*, caused by its two long roots. This shape led to the name zeuglodonts ("yoke-teeth") for these whales. Pennies (19 mm in diameter) for scale in all three drawings.

with Latinized, italicized words. The first part starts with a capital, and refers to the genus, such as *Homo* for our own genus. The second part of the name refers to the species, *sapiens*. So that makes *Homo sapiens*. The genus name is much like a human surname in that it is shared by relatives—by related species in the zoological case, or related individuals among people. Thus, all my family members are also "Thewissen," and the extinct cousins of my species are also *Homo*. The species name is much like a given name in people—many unrelated individuals might have the same given name, Hans in my case. In zoology, there is only one species that combines a particular genus name with a particular species indication. Zoological names are more strictly policed than the names of people; the Commission on Zoological Nomenclature evaluates conflicts and passes a binding judgment.

Groups of genera (plural for genus) are grouped in a subfamily, and several subfamilies into a family, and so on into more inclusive and larger groupings. There are also rules for how names for those groupings should be made. They are usually characterized by their endings in Latin. As an example, table 1 gives names for our own species (*Homo sapiens*) and a dolphin (*Delphinus communis*), and it also lists more and more inclusive groups into which these two species are categorized. Note how the endings of the words work. Delphinidae and Delphinoidea are very similar

TABLE I. EXAMPLES OF ZOOLOGICAL CLASSIFICATION.

Category	Typical ending	Modern humans		Common dolphins	
		Latin name	English name	Latin name	English name
Order		Primates	primates	Cetacea	cetaceans
Superfamily	oidea	Hominoidea	hominoid	Delphinoidea	delphinoid
Family	idea	Hominidae	hominid	Delphinidae	delphinid
Subfamily	inae	Homininae	hominine		
Genus		*Homo*		*Delphinus*	
Species		*Homo sapiens*	modern human	*Delphinus delphis*	common dolphin

words, but they mean different things: Delphinoidea includes all of Delphinidae, but also several other families (not mentioned in the table).

Neither commission nor rules existed in Owen's time, but later zoologists decided that the rules should be applied retroactively. Harlan had proposed the name *Basilosaurus*, a valid genus name predating Owen's *Zeuglodon*. Thus we correct the famous Professor Owen in favor of Harlan's original name, in spite of its erroneous connotation. However, Harlan did not propose a species indication, whereas Owen did. Hence, Harlan's genus name is now combined with Owen's species indication, and the animal is called *Basilosaurus cetoides*.

In spite of Owen's good work, the reptilian ghost of *Basilosaurus* lived on. In 1842, S.B. Buckley excavated a sixty-five-foot vertebral column with portions of head and forelimb on the same plantation where Harlan had found basilosaurid bones. Eventually, these ended up in Boston, where they were seen by Reverend J.G. Wood, who announced the startling inference that basilosaurids were swimming in the seas surrounding New England in the pages of the *Atlantic Monthly*.[11] Wood starts his essay by listing a number of examples of natural history phenomena that, although initially held to be untrue, were later confirmed by solid observations. Then he takes on the case of sea serpents, commenting that:

> *It is not very difficult to be witty about traveler's tales, and it is very easy to be sarcastic. . . . As long as an assertion cannot be proved, skepticism is triumphant.*

Wood discusses many, in his eyes, credible sightings of sea serpents, and goes into depth in the observation of a sea serpent living near Nahant in Massachusetts. The animal, or animals, were sighted multiple times by many different people between 1819 and 1875. One of the observers reported: "The head seemed somewhat like a horse, the portion of neck exhibited above the water was about two feet in length, and a little beyond the neck there were a series of protuberances, reaching a distance of eighty feet." Another stated: "More than once, it reared its head more than six feet out of the water, and made directly for one of the boats; the spray dashing over its neck, and the protuberances of the back glittering in the sun. But it never attacked a boat, and though it came near enough to frighten the rowers, it always turned sharply and retreated." The Boston Society of Natural History pursued the issue and interviewed the observers of the animal on one boat in 1875 in detail, and one of them even sketched the animal (figure 7). Wood interviews witnesses and deems the observations credible. All of them suggest a snake-like body, between 60 and 100 feet long, with forelimbs, not scaly, black on top, white underneath, having small or no teeth, and swimming by up-and-down movements. Wood discusses the options as to what this animal might be. It is not a large aquatic reptile, they are extinct; it is not a large version of the tropical sea snakes, they swim by means of lateral movements. He concludes that it must be a long snake-like member of the Cetacea, since it breaks the surface to breathe and swims by up-and-down movements of its body. Given that no cetaceans with snake-like bodies were known, he proposes that the sea serpent of Nahant Bay is a living specimen of *Zeuglodon,* whose bones he had seen in Boston. He compares the head, as sketched by the boaters, in detail—it matches that of *Zeuglodon* "if clothed in flesh and blood." He concludes that this whale is different from all known ones, in that it has its nasal opening closer to the tip of the snout and not on the forehead. He ends by saying that future boaters should not scare off or try to kill the animal with gunfire; instead they should harpoon it so that it could be reeled in and studied.

Wood did not uncritically accept fisherman's tales. He looked for consistencies and independent lines of evidence, like a scientist would. It is only in his conclusion that there are some sloppy leaps. There are certainly no basilosaurids living near Cape Cod. But we can pardon Wood; although he wrote after the *Origin of Species,* little was known about the age of the earth, and it was held to be much younger than we now know it is. So *Zeuglodon* seemed within the reach of time of the Nahant Serpent.

The tales of bones of enormous sea creatures in the American South attracted many, including the Englishman Charles Lyell, often considered the father of geology. Lyell commented that in his visit to Alabama

FIGURE 8. *Dorudon atrox*, an extinct basilosaurid whale that roamed the oceans thirty-four to forty-one million years ago. Fossils of basilosaurids were already known before the time of Darwin. Until the late 1900s, they remained the oldest whales for which full skeletons were known.

nearly 150 years thereafter (figure 8). So, what do we know about these whales that in name seem closer to reptiles than to modern Cetacea?

As intermediates go, basilosaurids are well along the evolutionary line toward modern whales: they have already adapted their bodies for life in the water and could not move around on land. And yet, they do retain a number of hints of their terrestrial ancestry, the most dramatic being their tiny hind limbs, complete with knees and toes. Scientists who have scrutinized their anatomy in detail have found a number of clues to ancestral whales.

Let's imagine that the Reverend Wood had been right, and *Basilosaurus* could be harpooned, corralled into a bay and captured, and then displayed in a large aquarium with glass walls. As we approach the tank, Harlan's serpent in the flesh looks like a snake, with a narrow, eel-like body propelling itself through the water with sinuous movements. It is clearly a fully aquatic animal. However, as we come closer, we see that the beast has no scales and has paddle-shaped forelimbs: flippers. This is no snake. The body is sleek, and has no constriction where the neck should be. Toward the tail, tiny hind limbs are visible, but they are too small to bear weight or help in swimming. It has been suggested that they were used in mating,[16] similarly to the claspers that male sharks use to help hold on to females when they copulate. At the end of the tail, we find the best evidence confirming our hunch that this was a whale: *Basilosaurus* had a horizontal tailfin, a fluke.

Because the animal is actually more than thirty million years old, we have only its skeleton and so we do not know whether it had fur, or sparse hair, or was naked like modern whales. Some scientists have tried to figure this out by studying modern animals, but the results remain ambiguous.[17]

Our information of *Basilosaurus, Zygorhiza,* and their relatives comes from the fossil skeletons that were mostly found in the deserts of Egypt and in the southern United States. These rocks were formed between thirty-four and forty-one million years ago. Both genera are included in the family Basilosauridae (basilosaurids in English),[18] which is traditionally divided into two subfamilies: Basilosaurinae, giant, elongated snake-like forms, and Dorudontinae (figure 8), shorter forms that somewhat resemble a dolphin in body shape.[19] In everything except their vertebrae and body shape, the two groups are very similar. Complete skeletons of the basilosaurine *Basilosaurus* show that the animal was about eighteen meters (sixty feet) long, whereas dorudontines, for instance *Dorudon,* were around a quarter of that (figure 9).[20] Basilosaurids have been discovered in many places all over the world and were probably distributed worldwide (figure 10).

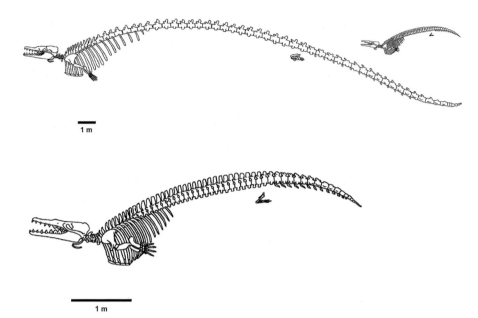

1 m

1 m

FIGURE 9. Skeletons of two fossil basilosaurid whales: the large *Basilosaurus* and the much smaller *Dorudon*. The picture of *Dorudon* is repeated in the upper right corner, but now at the same scale as *Basilosaurus,* to show the great difference in size. After Kellogg (1936), Gingerich et al. (1990), and Uhen (2004).

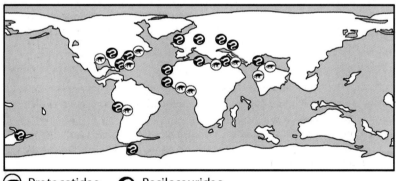

Protocetidae   Basilosauridae

FIGURE 10. Map of the world forty-five million years ago (the Eocene period), with places where fossils of basilosaurid whales and protocetid whales (discussed in chapter 12) have been found. Base map from http://www.searchanddiscovery.com /documents/2010/30109andrus/images/fig02lg.jpg and data points from http:// fossilworks.org where the cetacean sections are compiled and edited by Mark Uhen; he reports that the record from Antarctica is ambiguous.

Feeding and Diet. If our captive basilosaurid opened its mouth, it would be immediately clear that it was not like any other living whale. Most modern toothed whales have teeth that are simple pronged stabbers—think of the peg-like teeth of killer whales—with little variation across the tooth row or between upper and lower teeth. This similarity in tooth shape is called homodonty (*homoios* is Greek for "similar to"). But basilosaurids had more complex teeth that differ from front to back in the mouth, like humans and most other mammals. This is called heterodonty (*hetero,* Greek for "the other, different"). In the front, long and sturdy pointed teeth would be visible, whereas in the back, each tooth would have multiple bumps (or cusps, as paleontologists call them; figure 11).

Teeth and Paleontology

Teeth are very important to mammal paleontologists, because they are the elements that are most commonly preserved and because they are highly characteristic in different species. One tooth is often enough to identify a species. Owen identified *Basilosaurus* as a mammal based on a few teeth. Most mammals have four different kinds of teeth in each jaw quadrant—left and right upper, and left and right lower (figure 11). Think of your own teeth. From front to back, humans and most other mammals have incisors, canine, premolars (bicuspids), and molars. Primitive placental mammals, such as moles, have three incisors, one canine, four premolars, and three molars on both left and right in both the upper and lower jaw. Paleontologists express that as a dental formula 3.1.4.3/3.1.4.3, where half of the upper jaw is shown in front of the slash, and half of the lower jaw behind the slash. The dental formula is very stable within a species, but can vary greatly from the primitive placental count. In mice, for instance, the dental formula is 1.0.0.3/1.0.0.3. In humans, it is 2.1.2.3/2.1.2.3.[21] Often the upper and lower dental formulas are not the same. *Basilosaurus* is an example: 3.1.4.2/3.1.4.3. Thus, it has two upper molars but three lower molars. Throughout evolution, many mammalian groups independently have reduced the number of teeth from that original number, an important trend that we will return to in chapter 15.

Most mammals have relatively simple incisors (see Box), including *Basilosaurus,* which has a simple pointed prong with one root.[22] In most mammals the canine is bigger than the incisors, but in *Basilosaurus* it is similar to the incisors. The premolars of *Basilosaurus*

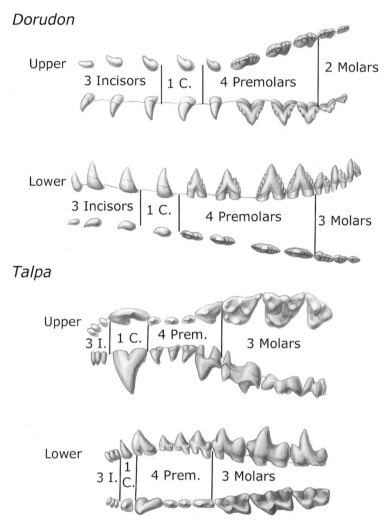

FIGURE 11. The adult dentition of the basilosaurid whale *Dorudon* and the mole *Talpa,* at very different scales. For each species there are side views of the left teeth (lateral views, middle drawings) and views of the chewing surface (occlusal views, top and bottom). *Talpa* has a dentition that is characteristic of basal placental mammals, including early ancestors of whales: three incisors, one canine, four premolars, and three molars on left and right side, with each of the molars showing complex morphology of lows and highs (cusps). *Dorudon* teeth are simpler in shape: the valleys between cusps have disappeared, and there are only two upper molars.

increase in numbers of cusps from front to back. Molars are also complex: each molar crown has a row of pointed cusps, and each molar has two roots.

A number of basilosaurid fossils are from young individuals who still had their milk, or baby, teeth. That is surprising, because modern cetaceans do not have a milk dentition: the first generation of teeth that erupt in a baby dolphin are the only teeth it will ever have. Thus, that too has changed in cetacean evolution.

The entire dental ensemble in basilosaurids was powered by very significant jaw muscles that covered the entire top of the skull, arising from a large crest on top of the head, the sagittal crest. No doubt about it: basilosaurids could bite hard. What did they eat? Some of the articulated skeletons show an accumulation of fish bones located in the area where the stomach would have been, and these have been interpreted as stomach contents.[23] Microscopic scratches on the teeth also look like the scratches in modern fish-eaters like seals,[24] so it seems that basilosaurids ate fish. One specimen has shark teeth in its belly, showing that small sharks at least were not a match for the King Lizard of Cape Cod. Also, there are tooth marks on the skull of a juvenile *Dorudon* that match the distance between teeth of *Basilosaurus*, suggesting that one basilosaurid ate others.[25]

Brain. Since the 1800s, a valley in Egypt called Zeuglodon Valley, or Wadi al-Hitan (Valley of the Whales), has yielded a wealth of basilosaurid skeletons. There, fierce winds scour the surface and carry away the sediment, exposing fossils. The exposure is temporary: eventually the fossil bone too is devoured by the wind, turned to powder, and blown away. Cavities in the fossils, such as the cavity in the skull where the brain used to be, are filled with fine sediment which is harder than the bone. So as the bone erodes away, the filled cavities remain. As a result, many fossils from this area are endocasts: lumps of hard sediment that preserve the shape of the cavity they once filled. Not only bone leaves impressions on the sediment. Many of the soft structures inside the braincase leave impressions on the bone too, making it possible to learn about anatomy that itself does not fossilize. Researchers have described cranial endocasts of basilosaurids in detail, and some were even named as separate species by overzealous paleontologists.[26] Endocasts can be used to estimate brain size, too. From this, it is clear that basilosaurids had tiny brains, much smaller than even those of modern cetaceans with small brains, such as bowhead whales (see Box).[27]

Brain Size

The volume of endocast of the cranial cavity (where the brain sits) can be measured and used as an estimate of brain size, and provide some indications of an ancient animal's intelligence. The cranial cavity contains several organs besides the brain, such as arteries, nerves, and the membranes that protect the brain (the meninges). Those structures often also leave impressions in an endocast, and those impressions are often not clearly distinct from the brain impression itself. So, measures of endocranial volume are an overestimate of actual brain volume in vertebrates. In a horse, for instance, 94 percent of the cranial cavity is filled by the brain.[28] Matters are worse in cetaceans, because, at least in the modern species, a large mass of veins envelops the brain. Such masses are called retia mirabilia (plural of rete mirabile or "wonder net"). Endocranial size has been estimated by dunking endocasts in water and seeing how much water is displaced, or, in modern times, by using CT-scan technology.[29] This has given us a good idea of how endocranial size changed in cetacean evolution. However, brain size may not follow this pattern, because the size of the rete may also have changed in evolution. Actual measurements on the skull and brain in a bowhead whale, a modern baleen whale, indicate that only 35 to 41 percent of the cranial cavity is filled with brain in this species.[30] That makes it difficult to tease apart the pattern of brain evolution from that of endocast evolution, although some broad patterns emerge.

Brain size is most meaningful when it is scaled with body size. Larger animals have larger brains simply because a larger body needs a larger brain to operate it. So if we are interested in studying brain size, we need to correct for body size. To make that comparison, scientists calculate a ratio called the encephalization quotient (EQ).[31] At any one body size, a mammal with an average-sized brain has an EQ of 1, an animal with a larger-than-average brain has an EQ greater than 1, and a smaller-than-average brain an EQ smaller than 1. Cats, for instance, have an EQ of 1; they have an average brain size for their body weight. Horses have an EQ of 0.9, and it is 2.5 in chimpanzees. Humans have the highest EQ on the planet: over 7. In the bowhead whale, the EQ is 0.4,[32] similar to that of a rabbit. The point has been made that this number is misleading since fat makes up 40 to 50 percent of the weight of a whale and fat needs less brain tissue to operate it than other tissues do, thus artificially lowering the EQ. If we correct the body-weight value by ignoring the weight caused by fat altogether, the recalculated EQ for a bowhead is 0.6, still low.

Vision, Smell, and Hearing. If you were watching our captive basilosaurid come up to breathe, you would probably notice that its nose opening was halfway between the tip of the snout and the eyes. It is unclear why the opening is so far back, although scientists have speculated that underwater life favored this position. After all, most living whales have blowholes far back on their heads, and can breathe while just exposing the smallest part of their body. But most vertebrates that live underwater have their nose opening at the tip of the snout, for instance seals, manatees, hippos, muskrats, and even underwater predators such as crocodiles, sea snakes, and sperm whales. There may be more to the evolution of the blowhole than just underwater living. The shifted position of the nasal opening certainly caused there to be less room for tissues involved with the sense of smell, but from the bones of the nose, it is clear that basilosaurids had a sense of smell.

The eyes of basilosaurids were directed toward the sides; they are located under a broad shelf in the skull, called the supraorbital shelf of the frontal bone. Their visual field is thus mostly directed toward the side, and this suggests that they were hunting prey underwater, which is consistent with what we know about their diet.

We know a lot about basilosaurid hearing, because many of their fossils are very well preserved and include such rarely preserved pieces as the ear ossicles (figure 3). Their ear ossicles are very similar to those of modern whales,[33] suggesting that, like modern whales, basilosaurids had keen hearing underwater (see chapter 11).

Walking and Swimming. With their serpentine body and tiny hind limbs, basilosaurids could not get around on land. Their home was the ocean—they are obligate aquatic animals. The vertebral column reveals that basilosaurids are mammals, not dinosaurs or fish. There are seven neck vertebrae, a number typical of mammals from giraffe to human. In basilosaurids, as well as modern whales, these vertebrae are very short; as a result, the shoulders are so close to the head that the neck disappears. Then there are seventeen thoracic (back) vertebrae, each of which carries a pair of ribs. Those ribs are interesting.[34] The part that reaches to the chest side of the animal (the ventral part) is very heavy and dense, a condition called osteosclerosis (*os* means bone in Latin; *scleros* means hard). This part of the ribs is also a bit thicker than the rest, which is called pachyostosis (*pachus* means fat in Greek). Such extra weight in the skeleton is important in some marine mammals because it provides ballast that allows them to stay submerged.[35] But these features are not usually present in fast predators such as many modern whales and dolphins, and compared to other mammals, basilosaurid bones are just mildly osteosclerotic.[36] It remains likely that dorudontines, like the dolphins that are similar in shape, were pursuit predators of fast-moving fishy prey.

However, the difference from modern whales begs for an explanation. Why do basilosaurids' ribs weigh them down? The position of the pachyostosis, on the ventral side of the rib, is suggestive. Perhaps, by concentrating weight on the belly side, the weight helped to keep the animal from going belly-up during swimming. The dorsal fin of modern whales—which is not made of bone and so wouldn't fossilize—helps with that job, acting like the keel on a ship in preventing rolling. We don't know if basilosaurids had a dorsal fin, but it may be that they lacked one and that the pachyostosis was an anti-rolling device.

Behind the thoracic vertebrae, *Dorudon* has forty-one vertebrae that change only very gradually in shape and size, reaching the tip of the tail. In land mammals, these vertebrae are divided into lumbar, sacral, and caudal vertebrae, and vary greatly in shape. The sacral vertebrae of land mammals fuse together to form a composite bone, the sacrum, that transmits weight to the pelvis and from there to the hind limbs.[37]

In modern whales, no vertebrae in this region fuse. Paleontologist Mark Uhen of George Mason University studied *Dorudon* in detail and found that even though there is no sacrum, vertebrae 17 through 20, behind the thoracic vertebrae, are different. These vertebrae have projections (transverse processes) that are much thicker than those of adjacent vertebrae. In land mammals, these transverse processes on the sacrum articulate with the pelvis, connecting hind limb to vertebral column.[38] It is likely that these vertebrae represent the sacrum. That allows us to identify the sacral vertebrae in this fossil whale, and the lumbar vertebrae in front of it. And that shows that there are many more lumbar vertebrae in basilosaurids than in most land mammals, although it is not clear what the function of this is.

In *Basilosaurus*, the vertebral numbers are similar to those of *Dorudon* but their shape is different. In *Basilosaurus*, the centrum of the lumbar and caudal vertebrae is enormous, like a massive cylinder larger than a can of paint; the vertebral arch is very small in comparison (figure 6). This allows for great mobility in all directions, as expected if the animal was snake-like in its locomotion.[39]

The triangular fin at the end of the tail of a modern whale is called the fluke, and basilosaurids had one too. In modern whales, the fluke is the shape of a symmetrical triangle.[40] Internally, the fluke consists of a row of tail vertebrae that runs down the center, with thick triangular

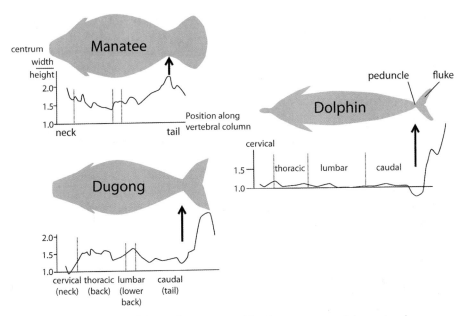

FIGURE 12. Vertebral shape of two seacows (sirenians: manatee and dugong) and a cetacean (the modern dolphin *Delphinus*). Note how in the animals with flukes (dugong and dolphin) the vertebrae are abruptly narrower at the point where the fluke is attached to the body (the peduncle). Flukes do not fossilize, but vertebral shape change can be used to infer the presence of a fluke in fossils. After Buchholtz (1998).

pads of connective tissue and skin making the extensions toward the sides. There is no bone in the triangular side flaps, so it does not fossilize and we don't have a preserved basilosaurid fluke. But we know they had one, because tail vertebrae in an animal with flukes are different from those in a non-fluked one. The massive part of these vertebrae (the centrum) has different proportions from the vertebrae in front and behind it (figure 12). The base of the fluke is called the peduncle, and here the centrum of the vertebrae is higher than it is wide, whereas further to the front and further to the back, these proportions are reversed.

In addition to that difference, the vertebra located right at the base of the fluke has convex anterior and posterior surfaces, and is called the ball vertebra. Basilosaurines and dorudontines had both these features, and so they had a fluke. It is likely that the dorudontines used their fluke to propel themselves during swimming, similarly to modern cetaceans. On the other hand, this is less clear in *Basilosaurus*. It is often said that these whales did not use their fluke for propulsion but swam by means of serpentine movements of their spine, as helped by the very flexible

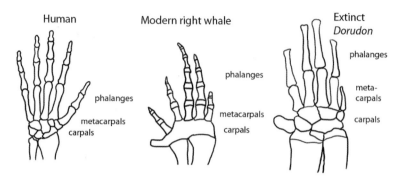

FIGURE 13. Hand of human and flipper of two whales. Humans show the ancestral pattern: five fingers, each with a metacarpal and three phalanges (two in the thumb). The fossil whale *Dorudon* has a single phalanx in each finger, or possibly one more phalanx in some individuals. Modern whales, such as the right whale, often have more phalanges, and the fingers are always embedded in a flipper.

spine. Body shape and swimming methods have been studied a lot in fish,[41] and *Basilosaurus*'s body has been compared to a giant eel, although some basilosaurid specialists, such as Uhen, doubt this interpretation.

Living cetaceans use their forelimbs mostly for steering, balance, and in starting and stopping; these limbs barely help in propulsion. Basilosaurids may have done the same. For the wrist and hand, only a few fragments were known. Mark Uhen described the forelimb for *Dorudon* and he found that the shoulder joint was relatively mobile, similar to that of a modern cetacean.[42] The elbow is not mobile in modern cetaceans, whereas that of basilosaurids allowed some bending and stretching. Movements at the wrist were just about impossible in basilosaurids, as in modern cetaceans. The fingers allowed some movements, unlike most modern cetaceans.[43] By comparing to other marine mammals, Uhen concluded that basilosaurid hands (or forefeet) were embedded in a stiff paddle, a flipper, just like modern cetaceans (figure 13). Inside that flipper were five bony fingers, as in most modern cetaceans as well as most other mammals. In humans, the palm of the hand holds five bones called metacarpals; each of those is followed by three bones (phalanges) that make up the segments of the fingers (the thumb has just two). In basilosaurids, the fingers appear to have had a metacarpal and just one phalanx, although it is not clear whether these bones were missing in the living animal or were lost during fossilization. If they were indeed absent in the living animal, it would mean that basilosaurids

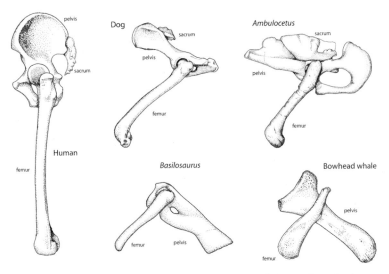

FIGURE 14. Sacrum, pelvis (innominate) and femur (thigh bone) of two land mammals, two fossil whales, and one modern whale. In most mammals, the sacrum consists of multiple vertebrae, one of which articulates with the pelvis, and the pelvis has a mobile joint with the femur (human, dog, and *Ambulocetus*, discussed in chapter 4). In *Basilosaurus* and all modern cetaceans, the connection to the vertebral column is lost. However, *Basilosaurus* still retains a joint between pelvis and femur.

lost, in evolution, two phalanges per finger compared to their ancestors, which would have interesting implications for their embryonic development (see chapter 13). It would also be surprising from an evolutionary perspective, since most modern whales have three or more phalanges. There are too many "ifs" here, but if some basilosaurids are ancestral to modern whales, the number of phalanges for the fingers would have gone from three (in the land ancestors of whales) to one or two (in basilosaurids) and back to three or more again (in most modern whales). More fossils are needed to clear up this issue.

Basilosaurus has tiny hind limbs, a few feet long, attached to a sixty-foot-long body. Although no complete hind limbs are known for basilosaurids, there are enough fossils to indicate that the other basilosaurids had hind limbs similar to *Basilosaurus*. The hind limb is attached to a pelvis, which, in land mammals, articulates with the sacrum (figure 14). To understand the hind limb of basilosaurids, it is useful to first consider the hind limb in modern cetaceans. Although the number and size of the hind-limb bones in modern cetaceans varies among species, it does not protrude from the body in any modern species: all bones are embedded in the wall of the abdomen (although we'll get to some exceptions in chapter 12). Bowhead whales have more parts to their hind limbs than most other modern cetaceans, although they vary in size among individuals. In bowheads, there is always a pelvis and femur, a cartilaginous or bony tibia, and sometimes even a bony metatarsal. Sometimes, there is a real synovial joint (a joint with lubricating fluid, like all of the highly mobile joints in the body).[44] The left and right pelvis of modern cetaceans do not articulate with each other, and they also do not articulate with the sacrum (figure 15). In many other modern cetaceans, there are no hind-limb bones, and the pelvis is a simple prong-shaped bone.[45]

Although it is not involved in locomotion, the pelvis of modern cetaceans does have a function. In the male, the pelvis anchors the muscles to the penis and to the abdominal muscles,[46] and muscles extend from these bones to the genitals in the female too.

The pelvis and femur of *Basilosaurus* were first described in 1900.[47] That fossil (figure 14) shows that there was a synovial joint between pelvis and femur and foramen behind it, like in land mammals, and unlike (nearly all) modern whales. Those features make it possible to determine how it was oriented in the body. One end of the bone is bumpy in texture, and has been interpreted as the point where left and right pelvis attach to each other in the body's midline (the pubic symphysis). However, modern bowhead whales have a similar textured area where the penis is anchored. It is likely that this is similar in basilosaurids, and that left and right pelvis of basilosaurids did not articulate with each other or with the sacrum. In shape of the pelvis, basilosaurids may be closer to modern whales than to other Eocene whales.

Most of the remainder of the foot is known from *Basilosaurus* from Egypt. *Basilosaurus* had a mobile knee with patella (kneecap), but the ankle consisted mostly of immobile and fused bones. The foot had three toes, and instead of a metatarsal and three phalanges, *Basilosaurus* had a metatarsal and just two phalanges, and those two fused into a single unit. Clearly, this animal could not bend its toes.

Habitat and Life History. We know little about basilosaurid social behavior. Some clues can come from the relative sizes of males and females. In some mammals (including many seals and sea lions, for instance, as well as gorillas), males are larger than females. Such a difference in size between the sexes usually occurs when males mate with multiple females each year (harems). In other marine mammals, most

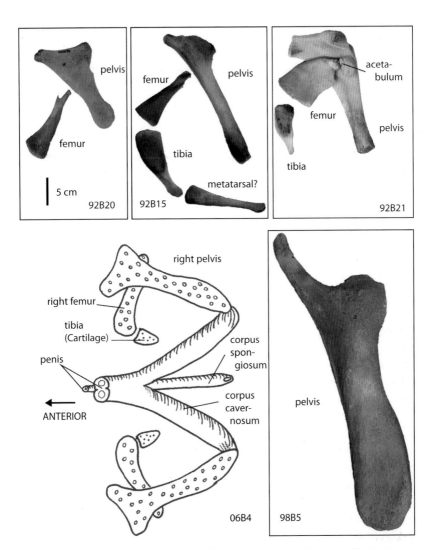

FIGURE 15. Pelvis and hind-limb bones in four modern bowhead whale individuals, showing the great variation in size and shape (the scale is the same for all four), with some having an acetabulum (joint for femur) and others a tibia and even a possible metatarsal. The diagram shows how these bones are oriented in a male bowhead whale as seen from the top (dorsal view).

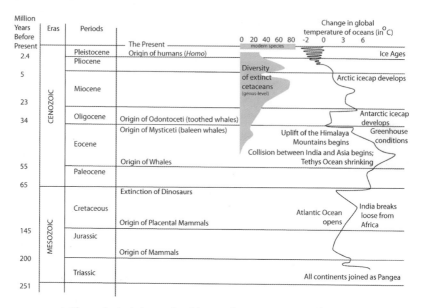

FIGURE 16. The geological time scale with some important events. Change in temperature patterns from Zachos et al. (2001); diversity of whales from F. G. Marx and M. D. Uhen, "Climate, Critters, and Cetaceans: Cenozoic Drivers of the Evolution of Modern Whales," (2010) *Science* 327 (2010): 993–96.

baleen whales for instance, females are larger than males, and in these species, males do not maintain a harem. The fossil record gives no indication that basilosaurid males looked different from females, so there were probably no basilosaurid harems.

Most specimens of basilosaurids have been found in rocks that indicate that they lived in a shallow sea,[48] but some species apparently preferred specific environments. For instance, the dorudontine *Saghacetus* is mostly found in sediments that indicate lagoons, whereas *Basilosaurus* is found in rocks that were formed in open water, away from the beach. Basilosaurids have been found in most oceans (figure 10), suggesting that they swam well enough to cross large seas. The climate in the time that basilosaurids lived, the late Eocene, was warm. The poles were bare of ice caps, and the temperature gradient from poles to equator was not nearly as pronounced as it is now. Near the end of the basilosaurids' reign, the planet changed (figure 16).[49] The continents shifted, which in turn transformed the oceanic currents, interrupting the mixing of equatorial and polar waters. As a result, the poles cooled, and at the end of the Eocene, Antarctica began to freeze over. Basilosaurids may have liked the more evenly warmed waters across the globe that occurred in the Eocene, and may have been unprepared for the sudden climatic cooling. Or they may have been out-competed by the new whales that were starting to show up on our planet, the ancestors of the modern toothed whales and baleen whales.

BASILOSAURIDS AND EVOLUTION

Basilosaurids are impressive cetaceans, resembling modern whales in many respects with their involucrum, flippers, and fluke. In some respects, they are intermediate between land mammals and modern whales. For example, the nasal opening is close to the forehead, unlike in land mammals, and the hind limbs were still present even though they were of no use in locomotion. The dentition is reminiscent of their land-mammal ancestors. For scientists, basilosaurids are intermediates: they are evidence that whales descended from terrestrial mammals. But basilosaurids look too much like modern whales to help us understand how that dramatic transition from land to sea happened. And they don't retain enough ancestral features to reveal just who their completely terrestrial ancestors were.

The paucity of the fossil record was fodder for those who doubt that evolution occurred and adhere to a Biblical account of the planet's history. Given the gap between four-footed mammals and basilosaurids, creationists pounced on whales as an example of the impossibility of evolution. Following up on the trouble Darwin had with whale origins, creationists claimed that no intermediates would ever be discovered. Alan Haywood wrote in 1985:

> Darwinists rarely mention the whale because it presents them with one of their most insoluble problems. They believe that somehow a whale must have evolved from an ordinary land-dwelling animal which took to the seas and lost its legs. . . . A land mammal that was in the process of becoming a whale would fall between two stools—it would not be fitted for life on land or at sea, and would have no hope of survival.[50]

For more than 150 years, basilosaurids were our best clue to what ancient whales looked like. In the early 1980s, West and Gingerich proposed that the Pakistani whales they found were older than basilosaurids and much closer to the land ancestors. But these Pakistani fossils were frustratingly incomplete. The new Pakistani incus suggested an intermediate condition for yet another organ system and added to the intrigue. But did it confirm the geographic region that should be scrutinized? Were there really undiscovered fossils, buried in Pakistan, that could bridge this gap?

CHAPTER 3

A Whale with Legs

THE BLACK AND WHITE HILLS

Punjab, Pakistan, December 1991. Last year's ill-fated field trip to
Pakistan has left me poor, so I can only afford to go alone to Pakistan
this time. There, Mr. Arif and I set out to do fieldwork in a blue Isuzu
pickup truck. That car was new in 1984, on my first trip to Pakistan.
Now, after eight years of off-road duty and poor maintenance, the car is
on its last legs. Every so often, Jamil, our lanky driver, pulls off to the
side of the road, and fiddles under the hood.

"Eek minute, sir, nooh problem."

The problem usually does go away, although in a bit more time than
Jamil's "one minute." Jamil's gray *shalwar kameez* has many oil stains.
It flaps in the wind as he moves his body energetically as if he is in his
twenties, but the creases in his face suggest at least another decade. He
knows and loves this car, but twice we interrupt the trip to visit a
mechanic. Next to car shops are usually tea shops, and we sip the sugary
brew of boiled milk and spiced tea while Jamil argues with two skinny,
brown, bare legs that stick out from underneath our car, not leaving the
car out of his sight for even a minute.

We arrive at the village of Thatta, just south of the Kala Chitta Hills.
There, a local politician owns a walled compound that resembles a
medieval village. Roughly hewn brown-gray rocks form a wall that
embraces a hilltop, and within it, separate walls surround five or six

houses with courtyards where his extended family lives. I recognize the rocks. They come from the mountains to the north and were formed in the Jurassic period, when a deep sea covered this area. The walls of each of the houses lack outward-facing windows, and there is only one small door to the outside, creating a feudal atmosphere ready to withstand the violent intruders of past ages. However, it is also functional in the present: the walls ensure the privacy of the women inside their own courtyard. They can uncover their faces here. The view over the outside wall looks down onto the village. It is a patchwork of tiny, one-room dwellings all built from that same rock type and all with tiny walled courtyards. Smoke fills the valley in the mornings when all the women bake bread in their clay ovens. There is only one paved intersection in this village, and barely any motorized traffic, so the sounds are those of kids playing, women banging cooking-pots, men yelling at their live-stock, dogs barking, birds singing at dawn, and five times a day the local *mullah* drowning out everybody else.

On the other side of our compound is the girls' school, on a hill and surrounded by a wall, but its buildings are modern, smooth and white-washed. The wall is high, so that the girls are afforded privacy too. But our hilltop is higher, so we can look over both walls and into their space. I avert my eyes when I am near, and Arif approves. It shows respect to the local culture.

The *mullah* wakes us around 5:30 A.M. with his call to prayer. It is broadcast through a loudspeaker, but none in our group gets up to pray. From most mosques, the call to prayer only lasts a few minutes, but this *mullah* also delivers what sounds like a sermon through his loudspeaker. It takes half an hour, and closes the book on going back to sleep for me. Arif chuckles from deep inside his sleeping bag.

"What is he saying, Arif?"

"He says that we people are worse than dogs, for not going to mosque."

"So most people don't pray in the morning?"

"Ahh, maybe they pray in house, maybe they do not pray. This poor man is a little bit sleepy."

Dogs are unclean animals in this culture, so that is quite an insult. But Mr. Arif thinks it funny, unconcerned as he is with the *mullah*'s empty mosque and his village ideas, and desiring to wake up slowly.

Life is simple. There is no heat, water, or gas, and it is very cold. I wear my coat to sleep in my sleeping bag. Our bedroom opens onto a veranda that opens onto the courtyard of our guesthouse. We are the only visitors. The kitchen has a single lightbulb as its only luxury. Rook-oon, the cook, brought a burner for cooking from the city. He fills it with gasoline; the *chappatis*, the Pakistani pancake-shaped bread, taste like fuel. We bring water in jugs from a well in the village, but it is contaminated, and all my companions have diarrhea. I do not, because I use a water filter.

"Water is bad. Jamil, Rookoon, this man, all ill, what you call?" Arif looks pained because he cannot remember the word as the four of us sit together wearing our coats in the dark kitchen.

"Diarrhea. I am not ill—I use a filter to clean the water. Shall I show you how to use it?"

I attach the hoses to the fist-sized filter, and pump the bad water from a glass through the filter into my empty mineral-water bottle, also brought from the city. Arif watches patiently, but not eagerly, as the bottle slowly fills.

I pour some water in his glass. He holds it up to the light of our bulb, swirls it as if it is fine wine, and drinks it.

"Taste is same," he says, unconvinced.

"You may use it anytime you like, so you can drink clean water."

Mr. Arif is silent. This is too far outside the sphere of things he knows. He does not take me up on the offer, and the diarrhea persists.

Walking through the kitchen into the courtyard, I have to bend my head because the door is too low. There is a bathroom in the compound too, a hole behind a wall, but there is no sewer system. With three companions with diarrhea, I stay away from it, holding my business until we are in the hills collecting fossils.

Jamil made a deal with a stray dog. He feeds her leftover *chappatis* every day. The dog, with its two puppies, now sleeps under his vehicle, and will bark if "bad people" come. I am not sure what Jamil expects, and I don't ask. Most Pakistanis, just like most other people, are somewhat embarrassed when a negative part of their country or culture surfaces. Jamil is a kind man. I do not want to embarrass him. Of course, with dogs being unclean, Jamil would not dream of touching the canine family. Not that the mom would let us get close to her—she waits until Jamil has walked away from the place where he drops the bread before she and her pups go and retrieve it. Our other pet is a peacock, which flies freely in and out of our courtyard. Arif lost his soap box, and he leaves his soap bar lying open on the veranda, where the peacock pecks at it. Every morning finds Arif searching for his soap, eventually locating it somewhere in the dirt, mangled by pecks and coated in sand. I do not know why the peacock likes soap, or why Arif never brings his bar inside to avoid the daily search and rescue.

Collecting fossils is easier this year. I now have a copy of a map made by a British geologist, T.G.B. Davis, before the country turned independent. Black lines indicate where different rock types can be found, and dashed lines where there are faults: cracks in the surface. There are a few topographical landmarks on the map that help to find places, although it does not show elevation.

Our field area is in the hills north of Thatta, less than half an hour from where we are staying. Fieldwork starts on January 2, and the next day, as we cross a low green hill covered with fossil clams, there is a rib fragment, as long as a finger but three times as thick. The bone is pachyostotic. It is unusual for a mammal to have thick ribs like that. The broken surface reveals that the small cavities that are normally present in bone are absent from this one: the bone is also osteosclerotic. Combining those two features, scientists call this bone structure pachyosteosclerotic.[1] Although basilosaurids are somewhat pachyostotic, there is only one group of modern marine mammals that has pachyosteosclerotic ribs: sirenians or seacows, which includes manatees and dugongs (figure 12). They are obligate marine mammals just like cetaceans, but unrelated to them; they became aquatic independently. The oysters have already indicated that these are rocks formed on a seafloor, so the rib pertains to a marine mammal, very likely a sirenian. Kind of cool to run into a marine mammal by accident. After last year's high with the *Pakicetus* incus, my interest in whales has waned. It is important to get back to land mammals, since that was what the grant money was for. I call Arif over, but he is not impressed. It is just a rib. He is right. I wrap my find, note it in my field notebook, and put it at the bottom of my backpack.

As the season continues, more fossils are found, but nothing spectacular. Apparently Richard Dehm, the German professor who worked here more than thirty years ago, picked up all the fossils that were on the surface and put them in his museum in Munich. Erosion is not fast here; not many new fossils have been exposed since that time. Like me, Dehm was interested in land mammals, and focused his energy on rocks that were formed in rivers. Geologically, we are in the foreland of the Himalayas. This area is very complex, and was greatly deformed when the mountains formed. Large slabs of fossil seafloor were pushed on top of river deposits, turned and twisted, and flipped on their sides. You always have to keep your eyes on the rocks here. A few steps and you may be out into a completely different fossil environment and millions of years later in time. The colors of these hills delight me. Mudstones are mouse-gray, dull purple, or the color of venous blood. There are white limestones in foot-thick bands that stick up higher than their surroundings and form ledges, so bright in the sun that it pains the eyes to look at them. There are silts, coarser than the mudstones; they are green, with lots of fossil clams, and show that the sea was there, fifty million years ago. The green color comes from the mineral glauconite. It forms in the wave zone along a coast. Its tiny crystals sparkle in the sun, as if a wandering giant crossed the green silts carrying a leaking bag of sugar. Together, these rocks are called the Kuldana Formation. They are visible in five low and nameless valleys, which I prosaically call A through E in my fieldbook. The rocks are occasionally overgrown by vegetation, gray-green thorny bushes that are widely spaced so that you can walk around them easily, and rocks are visible everywhere. The bushes make me feel as if I am in a park. I love this place.

In the distance are the higher hills that embrace all of valleys A through E. Those hills consist of sandstone, formed originally in rivers, millions of years after the coastal environment of the Kuldana Formation. They are black because of weathering, but cracking them with a hammer brings out their true color: brick red. Together, they form the Miocene Murree Formation, formed around thirty-five million years ago. The rocks are like a poor club sandwich, five thin sections of Kuldana cheese and cold-cuts separated by thick slices of Murree bread, and the entire thing set down on its side. Even farther to the north, beyond where I can see, are higher hills, made from light-gray limestones, formed in oceans more than seventy million years ago, when dinosaurs roamed the land. The Kala Chitta Hills have preserved their history: the story of an ocean that disappeared and was replaced by a massive river system (the precursor to the Indus), and the high mountains to the north. Even the name reflects the geology. *Kala chitta* is Punjabi for black and white: black for the Murree, white for the limestones.

People do live here, mostly herders with goats, sheep, and camels. We run into old men or boys who walk the flock, and Arif makes a point of chatting with them. Unable to speak the language, I continue to work.

The mudstones, mostly alternating gray and purple, are found in the older part of the Kuldana Formation (figure 17) and form the weathered floor of most valleys. However, occasionally among these mudstones there is a coarse, very hard layer, brick-red to purple in color. These are conglomerates, which look like layers of glued-together tiny pebbles. The pebbles vary in size from sugargrains to peas. Breaking a

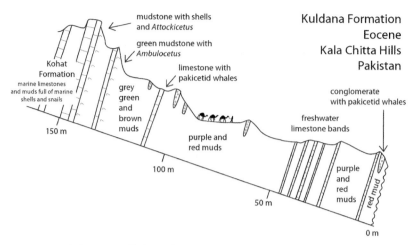

FIGURE 17. A diagram (called a geological section) of the layers that make up the Eocene of the Kala Chitta Hills in Pakistan. It shows the rocks that are found in a region in sequence and how thick each layer is. The tilt of the layers reflects their real tilt. Whales mentioned in this figure will be discussed in future chapters. Redrawn from S. M. Raza, "The Eocene Redbeds of the Kala Chitta Range (Northern Pakistan) and its Stratigraphic Implications on Himalayan Foredeep Basin," *Geological Bulletin of the University of Peshawar* 34 (2001): 83–104; L. N Cooper, J. G. M. Thewissen, and S. T. Hussain, "New Middle Eocene Archaeocetes (Cetacea:Mammalia) from the Kuldana Formation of northern Pakistan," *Journal of Vertebrate Paleontology* 29 (2009): 1289–98.

pebble, there may be bright whites and beiges, sometimes in bands, like a mini gumball. The pebbles are rounded lumps—geologists call them nodules. They form underground in hot, dry climates where groundwater evaporates. As the groundwater evaporates, the minerals dissolved in the water precipitate. These precipitates build layers, each mineral a different color, and form nodules. After the nodules were formed, a river came and washed away the mud that the nodules were formed in, collecting all the nodules and sweeping them downstream.[2] As the flow slowed, the river was unable to move the nodules and dropped all of them in what was probably an abandoned arm of the river channel. That arm may have held water long after the flow of the river stopped, forming a little lake. Subsequent cycles of rain, erosion, and mud deposition buried everything, including the bones of animals that died there. Finally, after the layers were deeply buried, groundwater percolated through them. The water carried calcium carbonate that precipitated out and glued the nodules together. Geologists usually carry a little bottle with strong acid. A drop of it on the calcium carbonate will dissolve it, and it will foam, like a shaken bottle of Sprite. Other rocks do not respond that way, making this a test for identifying calcium carbonate.

One of these hard conglomerates is fossil locality 62, and this is where the first fossil whale from Pakistan was found by Robert West, long before I ever worked here.[3] Locality 62 forms a low wall with fossils embedded in the hard rock. As we kneel down to inspect the fossils, we notice a braincase of a whale. It looks very similar to one Philip Gingerich collected across the Indus River from here, less than twenty miles away.[4] Philip called that whale *Pakicetus*. However, the skull is in the middle of conglomerate, which is as hard as concrete. I hammer the rock to see if I can extract the fossil by breaking the rock around it in a controlled way, but the rock does not budge. The fossil will certainly shatter if I continue. So, instead, I harden the fossil with glue to protect it from weathering and make a note in my fieldbook. This whale skull will remain alone for another year in the wilds of Pakistan, till I can return with heavier tools when I have money to come here again. Or rather, *if* I get money to come back here again.

A WALKING WHALE

A few days later, we sit down for our lunch: dry gasoline-scented *chappati*s with jam wrapped in newspaper, the newsprint readable in mirror image on the bread, because grease leached into it from the paper. Dessert is some nicely wrapped cookies from a store in Islamabad. We sit with our backs against a marine limestone that forms a ledge and comforts our backs, which hurt from carrying a backpack full of water and fossils and hunching over all morning. Sitting comfortably, I notice something blue against the gray rocks. It is a shark tooth, its enamel stained blue and white by erosion. This is not very important scientifically, but it brightens my day anyway, as fossils have been sparse today. As I put it away, there is another blue tooth, this time of a reef fish, with its heavy teeth and jaws, used to crush shellfish and clams. As I look up, there is a third fish tooth, all collected while I haven't even moved. It is puzzling to have all these vertebrates in marine rocks, while the river deposits are not nearly as fossiliferous. Should we spend more time in the marine rocks to collect fossils there?

We decide to work half-days on the fossil seafloor and the other half in freshwater rocks. As the fieldwork draws to a close, Jamil drops us in valley A, where there is a layer with lots of clam and snail shells braced by a blocky sandstone ridge of the Murree Formation. We walk along our mollusc bed, eyes to the ground. These molluscs lived in the Tethys Sea, a shallow sea that separated the Indian continent from Asia. Arif and I walk along the ridge, parallel to each other, keeping a distance so as to cover as much ground as possible.

Twenty minutes after we left the car, I find a distal femur, the knee-part of the thighbone, the first fossil of the day. It belongs to a beast the size of a cow, and clearly a mammal. The rocks indicate that this is a fossil seafloor, but I know of no marine mammal with knees like this. It is not a whale or a seacow; those have rudimentary knees or none at all. There is a group of mammals known from Pakistan that are thought to be closely related to elephants and seacows and maybe ancestral to them. No skeleton is known for these anthracobunids—this could be it. An exciting prospect.

I set down my backpack and crawl on my knees through the little valley. A thorny bush tears my shirt, and another one pulls off my hat and hooks my skin. Another fossil shows up: a proximal tibia, the other half of the knee joint. They are clearly from the same animal—perfect fit, same size, same sediment attached to them. This increases my excitement, because it shows that this was not just an isolated bone washing around in the ocean. The two bones stayed together, and I am hoping that they are part of an entire dead body that was fossilized here. Over time, two more pieces appear, both of the femur, but nothing nice. After forty-five minutes, Arif, who initially had slowed and was searching nearby so as to not get too far ahead, is moving on, eyes on the ground. An hour, and still nothing. I have to give up. I am disappointed. Nothing about this knee joint can tell me what this beast was. "Mammal," my fieldbook reports blandly, admitting defeat.

Arif calls me over. He has a green rock the size of a cereal box, with pieces of two bones. They share a joint and articulate as they would in the animal. I cringe. It is another knee, the proximal tibia and the distal femur, much smaller than the one I found earlier, and a painful reminder that I just wasted an hour. I try to suppress visions of working hard all day to come home with just two unidentifiable mammalian knees. I ask Arif where he found it. He points vaguely to the edge of the ridge he was walking on earlier. I do not want to deal with this now.

"Let's go on and keep working. We'll come back at the end of the day, on the way back to the car, and look for more of this beast."

Arif agrees. We walk on, eat our newspaper-wrapped lunch, and turn around in the mid-afternoon, toward the road, while covering new ground. Our bags are lighter now, because we've drunk most of the water we brought. Arif shows me the site of his knee. It is littered with fossil bone. It is immediately clear that this is much better than my site from this morning. There are ribs, phalanges, and pieces of larger bones, all in gray-green rock. This could be big. Maybe it is the first skeleton ever of an anthracobunid, and it may make it possible to study the relations between elephants and seacows. But there is no time to think about that now. It is time to start the excavation.

The bones are spread out over a small surface. This is a good sign: it means that erosion has not uncovered and disturbed a lot of the skeleton. First, we pick up all the loose bones at the surface, so that we do not trample them or cover them with dirt as we start to dig. The bones are weathering out of a green siltstone. It is hard, but not very hard. The layers are part of a vertical ledge. It is narrow and uncomfortable to stand here. As we inspect the site, we both take several involuntary slides down into the valley. After we've picked up all the loose fossils, the sun is about to drop behind the black Murree crest. This excavation will last many hours. The poor light forces us to go back to the road and head to the guesthouse.

The next day, we go immediately to the site, where I take notes. It is now locality 9209. I sketch rocks and layers in my fieldbook. We start to dig, and bones emerge immediately. There is a femur, of the opposite side from Arif's knee yesterday. That is great news, as it means that the fossil consists of more than just a single limb. We keep on digging, intensely concentrated. I try to keep notes, but it is hard not to get caught up in the excitement of the dig. We drip thin glue to harden fossils, and thick white glue to fill cracks. More notes, more excavation, waiting for more glue to dry. Our lunch is rushed; we want to dig. Two bones are in a block next to each other: radius and ulna, the bones between elbow and wrist. That is really exciting, a forelimb now, with wrist and hand bones, even. The rock is so hard that I need to take out the entire block and take it home, where I can use power tools to extract the fossils. That idea is scary, because I do not have the money to take extra luggage. But it is not a problem that I want to think about now.

Part of the fossil is articulated, and the positional information will help us to identify bones. I write numbers on each bone and draw maps in my fieldbook. My photos are poor, with irregular shadows covering the specimen. I am not a patient photographer.

We dig all day. Arif and I sit shoulder to shoulder on the ledge. It is addictive. Each new bone brings a high and makes us want more. We finish when it gets dark, exhausted but thoroughly satisfied. Conspicuous by its absence is the skull. The thought occurs to me that this is a beautiful skeleton without a head. That would make it very difficult to identify with certainty. Hopefully, the skull is still buried.

We come back the next day, and the process repeats itself. Halfway through the day, a herd of goats walks into the valley, long ears hanging down the sides of their heads, curved horns. They walk and eat, nibble the shrubs, and look in puzzlement at the two weird beings sitting in the dirt. They are not scared, and several times we have to shoo them away from our site. An old man is herding them. He is across from us in the valley, dirty blue checkered turban and long gray beard, and dressed in shirt and *dhoti*, a loin cloth that drapes down to his shoes. He carries a long walking stick that doubles as a club to steer recalcitrant goats. He comes near.

"*As-salamu alaykum*," Arif greets him.

"*Wa alaykumu s-salam*." He returns the greeting in a voice crackling with age.

Arif goes up the ledge to talk to him, but he walks closer. He looks at the bones we have excavated, picks some up and puts them down, and then walks over to me. His hand reaches to a bone still in the rock. He wants to pick it up.

"*Ney, ney*," I say emphatically, and he pulls his hand back, startled.

He and Arif talk for a while. I cannot understand him, and keep working. He asks the same question several times, and Arif answers it differently each time, patiently clarifying. Eventually, the shepherd leaves.

"Man is poor," says Arif.

"He has at least thirty goats."

"Ahh, goats are all goats from village, he is just herder."

That makes sense. They pool their goats for grazing purposes, so that it just takes a single person to deal with the day's grazing chore.

"What did he ask?"

"He asked if we find gold."

I set down my tools. This is bad news. If the villagers think we're finding gold, all kinds of things could happen. They could come and dig themselves, to share in the riches. They could ask us to give them some, or they could get the authorities to investigate, and it would take time to explain.

"What did you tell him?"

"I said, we people are government officers, doing important survey, no gold."

I don't reply. It seems like that statement would raise more suspicion, but I also trust Arif's judgment. The excavation continues, and Arif runs into a bone that is the length of his hand. With a dentist's scraper and a brush, I set to work to excavate it. Suddenly, I realize that this is the bottom edge of a lower jaw. There will be teeth, maybe even a skull. I will be able to identify this beast. I keep on working, not even telling Arif what this is, I am too tense. A black shiny surface protrudes above the dull tan bone—tooth enamel! I will know what this beast is! More of the skull is exposed, and it is embedded in hard green rock. No worries, the drill in the lab will take care of it, although it means that I will not get a good look at the teeth today.

We keep working at it, scraping and gluing. Eventually, we put the skull in a full plaster jacket, as if it were a broken leg, to protect it during transportation. The white plaster stands out from far away against the drab hills. We need to leave, because it is getting dark.

This worries me. I once collected fossils in a desolate desert miles from where people lived. We put a jacket on a lovely skeleton, and the next day, when we came back, the jacket had been pulled off and the bones scattered across the hill. It was inconceivable that in one evening and one early morning someone actually had walked by this out-of-the-way place, pulled off the jacket, and broken the fossil as well as my heart. The Kala Chitta Hills are much more populated. Someone certainly will see it here.

I consider covering the specimen with dirt. Arif argues against that. It will disrupt the rest of our excavation; there are still more bones buried. He is right. He pulls a sheet of paper out of my notebook, and writes, in the curly Arabic script that Urdu uses, "danger, explosive." He puts the paper on the jacket, and a rock on top of that to keep it down. The statement seems such an obviously silly lie that nobody will believe it. Arif disagrees, pointing out that local people do not use plaster, so they do not recognize the material, and that poor village folks respect authority. I again trust Arif's insight into the local people's psyche, and we leave the skull exposed. Still, I go to bed with a nagging feeling in my stomach.

We are back the next morning, and Arif was right: the cast is undisturbed, the note still on top. By mid-afternoon, all exposed fossils have been removed and packed and a map made of the position of the fossils (figure 18). We spend another hour digging, but do not find more fossils, and call it a day. We cover the site with loose rock to protect it from being trampled by herders and their flocks, and we return to the guesthouse.

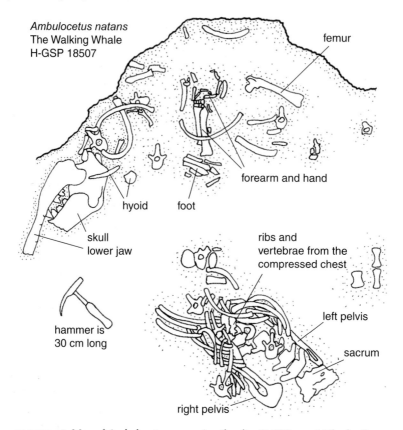

Ambulocetus natans
The Walking Whale
H-GSP 18507

femur

forearm and hand

hyoid   foot

skull
lower jaw

ribs and
vertebrae from the
compressed chest

left pelvis

sacrum

hammer is
30 cm long

right pelvis

FIGURE 18. Map of *Ambulocetus* excavation (locality H-GSP 9209). The fossil was in a nearly vertically oriented layer (see figure 19). The chest was deeper in the rock (lower on this drawing) and was not excavated until a few years after the initial finds. After S. I. Madar, J. G. M. Thewissen, and S. T. Hussain, "Additional Holotype Remains of Ambulocetus natans (Cetacea, Ambulocetidae), and Their Implications for Locomotion in Early Whales," *Journal of Vertebrate Paleontology* 22 (2002): 405–22.

It bothers me that it is not clear what kind of beast this is, and it also bothers me that I will have to wait until the skull comes out of the jacket to find out. It will be months from now. I try to numb my frustration by imagining that it is probably an anthracobunid. They must have had a heavy skeleton; their teeth are known from these rocks; and they lived in the ocean.

My frustration grows as I wash myself on the porch with a half a bucket of warm water, undressing only partially against the cold and to preserve modesty. The fossil bugs me. I think of myself as a patient person, but temptation is too strong here. Four days of work excavating this thing, and no resolution as to what it is. I decide to give in.

The numbers in the fieldbook identify which packages, wrapped in toilet paper, were found near the skull. I select two: the ear and the lower jaw. The floor of the veranda becomes littered with pink toilet paper as I unwrap them.

The ear part puzzles me. It is the size and shape of a small potato, and the bone is extremely dense, pachyosteosclerotic. One side is broken, but a thin lip of bone must have been attached here. There must have been a cavity there. The ears of elephants and sirenians look nothing like this. I feel that I should recognize it, but I do not.

Next comes the jaw. It is partly encased in rock, but some black tooth enamel is visible. I work on it with dental tool and toothbrush, exposing the side. It too fails to match my expectations. I am expecting the flat and squarish molar teeth of anthracobunids, but this tooth is exactly the opposite: high and triangular, with a second small triangular expansion behind. This is clearly not a sirenian relative. Then, what was it?

Suddenly, it hits me like a train. *Whales have teeth like that.* The potato-thing is the involucrum of the ear, dense as it should be.

This is a whale—a whale with, well, hind limbs. The first whale that walked. It is like a fog suddenly lifting, exposing a big city where there seemed to be nothing before. I sag back against a pillar of the porch with the jaw in my lap and the large orange setting sun stinging my face. We have discovered the skeleton of a whale that could walk and swim— the transitional form that paleontologists have wished for, and that creationists said would never be discovered.

I slowly recover. Dramatic intermediate forms are so rare in the fossil record that one really cannot count on ever finding one in one's lifetime. Understating the find, I write in the margins of my fieldbook for February 20:

> Decided that Arif's skeleton must be a whale (tooth and bulla). . . . There may be more in the wall, deep to what we have.[5]

It is clear that this is important and that more digging needs to be done. However, not now—the layer in which the fossil sits goes deeper, but it is difficult to access, and the field season is nearly over. The logistical predicament of my situation also forces itself on me: I am broke. Even if I dig all of it up, I will not be able to take it all home. As a matter of fact, I can't even bring everything home that I have right now. This skeleton needs three or more suitcases. Each extra suitcase means a fee of about $100, which I do not have. Moreover, it will take several years to get everything out of the rock, so even if it did come home, money would be needed to hire a professional fossil preparator.

I have to make a choice, and I choose the skull. That is the part that shows that this is a whale. The skull is also the part that is most difficult to prepare, and without it, no publication is possible. After the skull is out of the rock, the rest will be easy, and the excitement that it stirs may make it possible to get more money and come back here.

Back in Islamabad, I carefully pack the other remains in extra layers of newspaper and store them in two crates that used to hold oranges. Arif will safeguard them. The skull is swaddled in my dirty field clothes and goes in my suitcase.

Back in the United States, the work progresses slowly. I present my find in October of 1992 at a convention in Toronto that is attended by most vertebrate paleontologists. I am excited, and show lots of pictures of the skull, but I cannot show the hind limbs, I can only talk about them. Colleagues are polite but reserved. They buy the animal as a whale; after all, I can show them the teeth and the ears. However, without pictures of hands and hind limbs, they are unconvinced. True to their scientist nature, they are skeptical, reserving judgment until the bones can be viewed. The next year, Philip Gingerich, who collects in central Pakistan, offers to bring my orange crates back for me if I give him a sneak preview of what is inside. I gratefully accept, and the skull is reunited with its feet.

Finally, in 1994, all is ready and the beast can be presented to the scientific community and the public.[6] I also get to give it a name (figure 19). The animal represents a new genus and species, and is so different from all other whales that it is a new *family* of whales, too: *Ambulocetus natans*, in the family Ambulocetidae. The genus name describes what is most unusual about this fossil: it is a whale that walked. *Ambulare* is Latin for walking, and *natans* means swimming. It is the walking and swimming whale. In the week that my article is published, I spend most of my days talking to journalists about the find and its importance. I am not ready for all the press attention—my early interviews are clumsy—but the excitement it stirs up is exhilarating.

Scientist colleagues are excited now, too. Stephen Jay Gould devotes an essay to the find.[7] He writes, "If you had given me a blank piece of paper and a blank check, I could not have drawn you a theoretical intermediate any better or more convincing than *Ambulocetus*." *Discover* magazine

FIGURE 19. Life reconstruction of the fossil whale *Ambulocetus natans*, which lived in what is now northern Pakistan approximately forty-eight million years ago. *Ambulocetus* spent most of its life in water, but was able to come onto land, too.

includes whale origins in its top science stories of 1994. *Ambulocetus* opens the door to the recognition that the origin of whales is indeed documented in the fossil record. It is an exception to the common wisdom that transitional forms are difficult to find. I am excited about the opportunity to study how organ systems were transformed as whales evolved to become aquatic, from land to water. The first system I want to study is locomotion.

CHAPTER 4

Learning to Swim

MEETING THE KILLER WHALE

Stephen Jay Gould's essay in *Natural History*[1] highlighted one phrase in the article describing *Ambulocetus*: the phrase "the feet are enormous." He liked it because it cut through jargon and expressed some excitement. Indeed, *Ambulocetus*'s hind feet are as big as clown shoes, presumably because they become powerful oars in the water. The hands (or forefeet) are much smaller. In modern days, seals have feet bigger than their hands[2] because they use the former for propulsion when swimming, not the latter.[3] But seals and whales are not related, and all modern whales swim with their tails, so it is surprising that *Ambulocetus* had large feet. Also, in true seals (Phocidae in Latin), the feet move side to side in swimming, whereas a whale's tail moves up and down. Whales descended from quadrupedal (four-footed) land mammals, and that implies that their propulsive organ changed from limbs to tail. *Ambulocetus* showed that the feet were important in swimming, and thus, foot-propelled swimming came before tail-propelled swimming. But that leaves the question as to how those feet moved—was it up and down, like the tail in a modern whale, or side to side, like the feet in a swimming seal?

The fossils only go so far regarding those kinds of questions. Instead, one has to understand swimming in living mammals. I contact Frank Fish, who has studied swimming in mammals for most of his life. Frank is an avid swimmer himself, and, incidentally, knows all the jokes that people make linking his name with his field of research. Frank puts animals in a flow tank, which is an aquarium or pool where he can change the water flow, and films them swimming. Then he analyzes their movements in slow motion at different flow speeds, and applies his engineering knowledge to understand why which parts move. Musk-rats, for instance, swim by paddling with their feet. Their tail is flat from side to side, and it moves through the water like a corkscrew that balances the animal but contributes little to propulsion.[4] Frank's tank is too small for big mammals, so those he studies in marine parks. Frank is intrigued about *Ambulocetus* and invites me to come out and see his operation filming killer whales in a marine aquarium.

The filming is done early in the morning, before the park opens to the public. The trainer opens a door so we can go behind the scenes. As I walk in, a large black head suddenly emerges from the holding pen next to me and its eye stares directly at me. The killer whale has realized that we are not his trainers and caretakers, and checks us out. I am not used to having such a large living animal so close. It is unsettling.

Frank sets up cameras on long extension poles and arranges ladders to stand on while the trainers play with the whales. When they are ready, Frank mans the camera, shouting requests to the trainer.

"Now I want him to come full speed right underneath the camera."

With hand and sound signals, the trainer transmits the command, and the whale obliges.

"He just turned a bit when he was under the camera, can we do that again?"

I just stand around, absorbing the scene. The whales seem happy with the attention. This routine is different from what they usually do, and they appear eager to be part of it. As a matter of fact, one of the whales not involved in Frank's movie is looking over the wall between the two tanks. His trainer does not want him to feel ignored and throws a fish. The whale dives down and picks up a yellow maple leaf from the bottom of his tank. He sticks out his tongue to the trainer with the leaf on it. The trainer takes the leaf and throws it back in the water. A game of fetch starts. The trainer gently tugs the whale's tongue. The whale pulls back, but immediately sticks its tongue out again. Killer whales like having their tongue massaged.

FROM DOG-PADDLE TO TORPEDO

Frank has studied many whale and dolphin species, and they all swim similarly. When they propel themselves in a straight line, whales and dolphins use their tail, not their forelimbs (the flippers).[5] The fluke is pushed through the water, up and down, and both the upstroke and the downstroke help propel the whale. That is unlike swimming in humans. When humans do the breaststroke, the part of the stroke that closes the legs provides propulsion. It is called the power stroke. The rest of one cycle of the legs is the recovery stroke; it does not help with propulsion but just brings the legs back into position to be able to initiate another power stroke. A swimmer's speed falls during the recovery stroke. In the movement of whale flukes, there is no recovery stroke. It is obviously a much more efficient way to move, and similar to flapping bird wings[6] and fish tails,[7] even though those move very differently from flukes. Engineers call the force that moves the animal the lift force, and the surfaces that make lift (feet in seals or tail in cetaceans) are called hydrofoils. A special shape makes it possible to reorient the hydrofoil in such a way that propulsion is generated throughout the cycle. The movement through the water is complex, too. Hydrofoils differ in this way from paddles such as the oars of a rowboat or the feet of a human doing the breaststroke.[8]

Frank refers to the whales' mode of locomotion as caudal oscillation, because the tail is the hydrofoil (*cauda* is Latin for tail) and it swings back and forth. Most of the movement occurs in one area at the root of the tail, right where the ball vertebra is located, and known to exist in basilosaurids (see chapter 2). It works much like the hinge of a door.

A whole new world opens for me as I help Frank. My previous insights into locomotion came from the perspective of the boxes full of bones in museums and labs. That perspective leads to insights. It makes sense that seals have short legs with large feet. They can make short but powerful strokes, which is good for moving in a dense medium like water, but bad for land locomotion. But Frank's way of looking at the whole animal adds the actual movement, a new dimension.

Frank's work shows that mammals swim in very different ways. Whales and dolphins, and also manatees and dugongs (figure 12), swim by caudal oscillation when they go in a straight line. They keep their body stiff, streamlined like a torpedo. Seals are pelvic oscillators: their hind limbs move through the water side to side, without involvement from the tail.[9] Sea lions drag the back of their bodies when swimming, and are propelled by their large, wing-like forelimbs. The movements of those forelimbs resemble the wing beat of a bird;[10] that mode of locomotion is called pectoral oscillation. Cetaceans, sirenians (seacows), seals, and sea lions are the most aquatic mammals, but there are many other mammals that are good swimmers. Polar bears and some moles drag their hind limbs and paddle with the forelimbs (pectoral paddling), whereas beavers hold their forelimbs close to the body and paddle with their hind limbs (pelvic paddling).[11] There is a diverse world of swimmers out there that should help us understand why the fossils of past swimmers looked the way they looked.

Frank had thought about evolution, too, and after collecting data on lots of swimming mammals, he put it all together (figure 20), proposing how more efficient ways of swimming evolved from less efficient ones.[12] For the caudal oscillation of whales, understanding the swimming modes of otters and their relatives proved to be key.

The otters are in the same family of carnivorous mammals as committed landlubbers such as badgers, skunks, and wolverines, and that family also includes sleek-bodied weasels and martens. The otters all look similar in form—long and narrow bodies with short legs—but their extremities are very different. River otters have a short but relatively muscular tail and limbs. Sea otters are large-bodied, and they have very large, asymmetrical hind feet, with the little toe much longer than any of the others, and a little stub-like tail. Finally, in South America lives the giant freshwater otter, *Pteronura brasiliensis*. It is as big as a sea otter, but very different in shape, with small feet and a long and powerful tail, which is flat from top to bottom. All those differences in feet and tails relate to how these animals swim.

Minks, for instance, are a close terrestrial relative of otters, and are probably similar to the ancestral otter from the time before they were aquatic. Minks are land animals, but occasionally they do swim. Their long, sleek body is well suited to dashing through the underbrush without getting caught in branches. But minks are slow swimmers. They paddle with all four feet;[13] left fore and right rear beat at the same time, as the animal struggles to keep its head above water in order to breathe. River otters are different. In the zoo, one can see them dash unpredictably left and right and up and down with a real or imagined playmate, but that is not the kind of swimming that Frank can study. To measure joint movements and swimming speed precisely, he needs to see the animals go in a straight line. River otters swim using different strokes, depending on how fast they want to go.[14] They move from quadrupedal paddling and paddling with just the hind limbs (pelvic paddling), when swimming at the surface, to dorsoventral undulation at faster speeds underwater. The latter is the most efficient. Waves traveling through the vertebral column propel mostly the tail, but also the hind feet. The sea

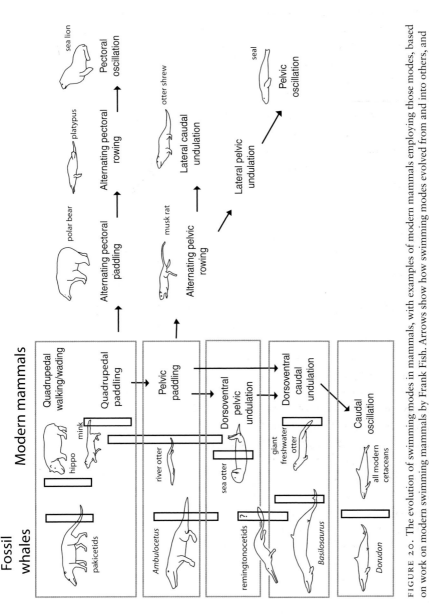

FIGURE 20. The evolution of swimming modes in mammals, with examples of modern mammals employing those modes, based on work on modern swimming mammals by Frank Fish. Arrows show how swimming modes evolved from and into others, and outlines show animals employing those modes. Boxes contain animals that swim similarly, mostly otter relatives and modern cetaceans. Clear bars indicate that some swimmers employ more than one swimming mode. Mustelids (otters and their relatives) are good models for the swimming evolution of whales, and their body proportions were used to infer the swimming modes of extinct cetaceans. Some of the fossil whales in this figure will be discussed in future chapters. Some drawings by Linda Spurlock.

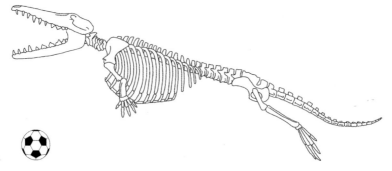

FIGURE 21. Skeleton of *Ambulocetus*, a forty-eight-million-year-old whale from Pakistan, based on Thewissen et al. (1996). Soccer ball is 22 cm (8.5 inches) in diameter.

otter swims underwater by moving its feet up and down, propelled by sinuous movements of its body: pelvic undulation. The feet are enormous and asymmetrical: they provide lift. Most interesting from a whale perspective is that giant South American freshwater otter. It propels itself with its long tail, which it swings through the water in an up-and-down fashion: caudal undulation. Frank put it all together and proposed that whales went through locomotor changes in their evolution that are mirrored in the members of the modern otter group. And he did this before any fossils documenting that transition were found.

That made the fossils the perfect way to check his results. If Frank was right, then the locomotor skeleton of *Ambulocetus* should match that of one of those otters. And indeed, *Ambulocetus* is proportionally like a river otter.[15] It is likely that the terrestrial ancestors of whales were quadrupedal paddlers, since most land mammals swim that way. From there, it is likely that swimming modes in whales changed a number of times, going through stages represented by modern otters—alternating pelvic paddling; simultaneous pelvic paddling and dorsoventral pelvic undulation; caudal undulation—to finally end up as caudal oscillators.

Since that work, more fossil whales have been discovered. An Eocene whale from India, *Kutchicetus* (discussed in detail in chapter 8), is younger than *Ambulocetus* and has flat tail vertebrae and short limbs, suggesting that it was a caudal undulator.[16] Other analyses elaborated on this work. A complex mathematical analysis of whale skeletons geologically younger than *Ambulocetus* confirmed that a hind-limb-dominated phase of swimming preceded the tail-based phases.[17] However, the results of that study did not find a link to mustelids, probably

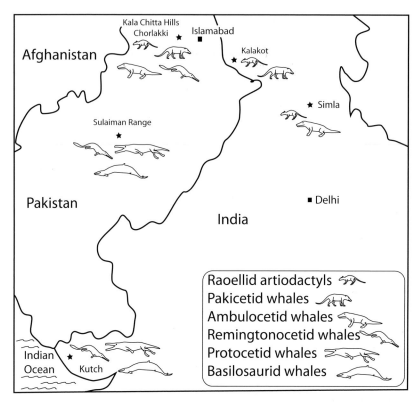

FIGURE 22. Sites where fossils of Eocene whales and raoellid artiodactyls have been found in Pakistan and western India. Pakicetid, ambulocetid, and remingtonocetid (see chapter 8) whales, as well as raoellid artiodactyls (see chapter 14), are only known from this part of the world.

because it did not include any data on the tail, which is probably of importance because it is the propulsive organ of modern whales.

We already saw that the fluke of *Dorudon* indicates that it was a caudal oscillator. *Basilosaurus* does not have an analogue among the otters. Even though it retained the fluke of its ancestors, its vertebral column was extremely flexible. It was probably an undulator,[18] along the lines of snakes and eels and different from any other cetacean.

That killer whale that looked at me as I entered its enclosure could hardly be more different from the otters that Frank studied: screamingly black and white, smooth and robotic, the size of small bus. However, when making its rounds through the water, the differences blur. Both otters and whales are perfectly at ease in water—gracious, fast, and acrobatic. The up-and-down movement is apparent in both, even though one has a fluke

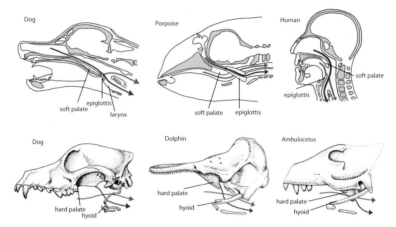

FIGURE 24. The paths of food (red) and air (blue) cross in the throat of mammals. Top diagrams show sections through the midline plane. Note how the red arrow passes to the side (laterally) of the blue one in all, but how different the relative location of soft palate and epiglottis is. Bottom drawings show the same paths superimposed on the skulls and hyoids of three mammals, showing the extension of *Ambulocetus*'s palate to the back.

Swallowing

The most puzzling part of the skull of *Ambulocetus* is the area of the throat. In most mammals, the bony part of the palate (the hard palate) ends near the back of the teeth (figure 24). Behind that is the soft palate, a wall of connective tissue and muscle that separates the rear of the mouth (the oral cavity) from the rear of the nose (the nasopharyngeal duct) just before both open into the throat.[22] Food is carried from mouth to the throat, and air is transported through the nasopharyngeal duct to the throat. In humans, a little tissue flap hangs from the back of the soft palate and is featured in many comic strips, but most animals do not have it. The throat anatomy of *Ambulocetus* is different from that of most mammals. The hard palate goes back much beyond the teeth, all the way to the ears, and the nasopharyngeal duct and hard palate flare down (ventrally), into the back of the oral cavity. The soft palate does not fossilize, so we do not know about its anatomy, but certainly mouth and nose were separated by bone much farther back than in most mammals.

Areas deeper in the neck are also different. In most land mammals, but not humans, the end of the soft palate touches a piece of cartilage that forms a valve (the epiglottis), and that contact helps to prevent choking. Closing that valve seals off the entry to the trachea (the windpipe) when the animal swallows, thus avoiding food going into the trachea.

In humans, the larynx is located lower in the neck, and the epiglottis does not reach the soft palate. Lacking the palate–epiglottis seal, humans choke much more easily than most mammals. However, the increased space between these structures is important in speech.

The human larynx moved down into the neck in evolution, whereas in odontocetes it has moved up,[23] and now protrudes into the nasopharyngeal duct, making a very tight seal. This causes the air passage and the food passage to be completely separated. No food or drink can go up the nose in a dolphin, important for an animal that feeds underwater. It is possible that the extended hard palate of *Ambulocetus* also served to keep food and air passages separate. However, if that was indeed its function, the mechanism was certainly very different from the mechanism that modern cetaceans use. More study is needed to understand the function of the nasopharyngeal duct in *Ambulocetus*, and it is not difficult to imagine other functions. Maybe, just like in humans, the throat was used in making sounds.

Even more puzzling are the bones that surround the throat of *Ambulocetus*. In all mammals, the larynx is supported by some bones and pieces of cartilage that together are called the hyoid.[24] The bones of the hyoid are not usually preserved in fossils, but in whales they are unusually large, and hyoid bones are known for a number of extinct species. In *Ambulocetus*, the three bones that make up the hyoid were found still partly in their position with regard to the skull (figure 18). However, their depth is barely larger than the depth of the nasopharyngeal duct. And yet, all food has to pass between this duct and the hyoid on its way to the stomach. The space available may just admit something the size of a golf ball, meaning that *Ambulocetus* only swallowed things smaller than that. Modern toothed whales do not chew their food. They tend to swallow large chunks. Killer whales are routinely found with entire seals in their stomachs, for instance.[25] *Ambulocetus*'s mouth and throat hardware clearly worked differently from modern whales—it ate smaller chunks—but more research is needed to fully understand the similarities and differences.

Vision and Hearing. The position of the eyes in *Ambulocetus* is different from that in basilosaurids. The eyes of *Basilosaurus* and *Dorudon* are large. They face laterally, and they are located on the side of the head, under the thick bony shelf called the supraorbital process. In *Ambulocetus*, the eyes are also large, but they are perched on top of the head, close to the midline, and they face partly sideways and partly up.[26] That position suggests that *Ambulocetus* could stay submerged while lifting just its eyes out of the water to survey its aerial surroundings, much as alligators do. *Ambulocetus* clearly had an interest in watching

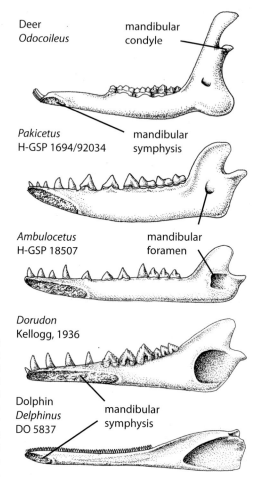

Deer
Odocoileus
mandibular condyle

Pakicetus
H-GSP 1694/92034
mandibular symphysis

Ambulocetus
H-GSP 18507
mandibular foramen

Dorudon
Kellogg, 1936

Dolphin
Delphinus
DO 5837
mandibular symphysis

FIGURE 25. The lower jaw of a deer and some fossil and modern cetaceans. The size of the mandibular foramen increases in whale evolution. This foramen is involved in sound transmission under water. The mandibular symphysis is very large in late Eocene whales (*Dorudon*). These drawings are not to the same scale.

things above the water, and it is possible that the whale found its prey there.

The shape of the lower jaw of *Ambulocetus* provides some clues about the evolution of hearing. In all mammals, there is a small hole in the back of the lower jaw, the mandibular foramen (figure 25). Through this foramen travel an artery, nerve, and vein that supply the lower teeth. Dentists target this nerve with their syringe if they need to numb the lower teeth of a patient. In most mammals, this foramen is just big enough to transmit those three structures. Not so in modern dolphins, other toothed whales, and basilosaurids, where the foramen is very large and houses a fat pad. That fat pad has an important function in hearing.[27] In *Ambulocetus*, as well as in *Himalayacetus*, the foramen is intermediate in size, larger than in land mammals but not the size of basilosaurids' and odontocetes' either.[28] That intermediate size suggests that the sound-transmission mechanism of the ear is evolving in these whales; this will be further discussed in chapter 11.

Walking and Swimming. Most of the vertebrae of the chest and back were found for that one individual of *Ambulocetus* initially found by Mr. Arif.[29] The neck vertebrae for this animal are poorly preserved, and much of the tail is missing. There are sixteen thoracic vertebrae with many preserved ribs, eight lumbar vertebrae, and four vertebrae fused into a sacrum. Many of these were found in articulation, so they were buried with flesh still holding them together.

These numbers are surprising. In most mammals, as well as birds and reptiles, the total number of cervical (neck), thoracic (chest), and lumbar (back) vertebrae is around twenty-six.[30] In birds, there are many cervical vertebrae and few lumbar vertebrae, but it still adds up to around twenty-six. In mammals, there are (nearly always) seven cervical vertebrae, and the thoracic and lumbar numbers vary inversely, making the total number of presacral vertebrae also add up to around twenty-six. *Ambulocetus*'s thirty-one is different, and basilosaurids have even more. Clearly, cetaceans are altering some of the basic structural designs of mammals. We will get back to that in chapter 12.

The *Ambulocetus* skeleton that we found was from a young individual. Many of its vertebrae still have areas where growth was taking place. In life, these areas contained cartilage discs called growth plates (epiphyseal plates), which disappear after growth ceases in an adult. In most cetaceans, growth plates disappear first in the neck and tail, and then move toward the vertebrae more or less in the middle of the animal. Indeed, this is also the case in *Ambulocetus*.[31]

Finding the sacrum of *Ambulocetus* was a real treat. It has four fused vertebrae, with a firm joint for the pelvis, similar to land mammals and different from other whales, including basilosaurids. The posterior part of the pelvis, where the hamstring muscles attach, is also large. The same is true in seals, where those muscles are used in kicking back the legs.

The forelimb was less flexible than in many land mammals: the radius and ulna (the bones between elbow and wrist) were immovably locked against each other. That indicates that the animal could not cup its hand (supinate). There were five fingers, with a normal complement

The fossil whale
Ambulocetus natans
Hand

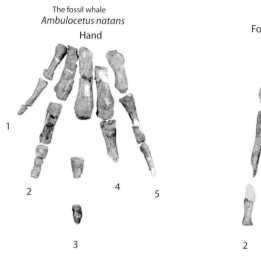

Foot

FIGURE 26. The bones of the hand (forefoot) and foot (hind foot) of *Ambulocetus natans* (H-GSP 18507). The animal had five fingers (1 is the thumb) but only four toes (labeled 2 through 5); it did not have a big toe. Gaps indicate bones that were not discovered.

of metacarpals and phalanges (figure 26). Fingers as well as toes ended in a low and long hoof, similar to the hoof of a deer. That suggests that cetaceans were related to hoofed mammals. Flanges on the sides of *Ambulocetus*'s phalanges of the foot indicate that the foot may have had webbed toes.

Habitat and Life History. With only one relatively complete specimen, not much can be said about the life history of the species. But even though this was a young individual, there was extreme and unusual tooth wear, suggesting heavy use. Given that there is a lot of relief between the front and back of each tooth, and that the tips of the molars are high, it is unlikely that some very abrasive food caused the wear. It is more likely that *Ambulocetus* chewed in a very specialized way, but we do not know how.

Shells of marine snails and sirenian ribs occur near the site where *Ambulocetus* was found, indicating that the ocean was near. *Ambulocetus* lived a hot climate, in water that varied in salinity.[32] However, freshwater was not far off, as evidenced by the fossils of land mammals in nearby rocks. It is likely that *Ambulocetus* lived on the edge between land and water, as well as on the edge between freshwater and saltwater. It was a transitional form in more than one way.

AMBULOCETUS AND EVOLUTION

Ambulocetus is often referred to as a missing link: a critical fossil that combines features of two groups that it is related to, while not much else ties those two groups together. From the perspective of creationists, missing links do not exist. Duane Gish devoted an issue of *Creation* to the beast and stated that it is "probably an animal related to seals."[33] With that, he acknowledged its amphibious nature, but ignored the plethora of cetacean features. Gish claimed that missing links cannot exist because they would have no hope of survival in between two different environments. When it was first described, *Ambulocetus* defused that argument, to the delight of the staunch defenders of evolution.[34] Such victories aside, it is important to remain humble about our understanding of all evolutionary transitions. As one colleague put it, "Every missing link that is discovered creates two new ones, one on either side of discovery."

Ambulocetus is a critical piece in the puzzle of cetacean origins, but most of the puzzle cannot be discerned. With this new fossil, many parts of the body are similar enough to land mammals that detailed comparisons can be made that may allow us to determine to which land mammal cetaceans are most closely related. It might contribute to a controversy that is brewing in the scientific world. Paleontologists assume that cetaceans are closely related to an extinct group of hoofed carnivorous mammals called mesonychians,[35] but molecular biologists note many similarities between the DNA and proteins of cetaceans and those of even-toed ungulates (artiodactyls) in particular hippopotamids.[36] But *Ambulocetus* does not resolve the question. More fossils of the first whales are needed, like those of that enigmatic *Pakicetus*. Pakistan is where both of those were found, making me think that the place of origin of cetaceans is in that region. I want to go back there, and find fossils that answer these questions. That is, if I get the funds, and if it remains safe to travel there. Those are big "ifs."

CHAPTER 5

When the Mountains Grew

THE HIGH HIMALAYAS

Plane Over Pakistan, May 23, 1994. Visits to the Indian subcontinent are best done between December and April, after it has recovered from the drenching monsoon rains in fall, but before the summer sun parches it. This year, I do not follow that recommendation. I arrive on the plains of Punjab during the mango season—the one benefit of traveling in May. I have to travel at this time because we want to go to the high Himalayas, where snow, avalanches, mudslides, and bitter cold make collecting in winter all but impossible. But the Indus Plain, where we will soon land, is miserable now. When the British ruled India, they moved the top level of government into the mountains during summer, leaving typhus and dysentery to rule the plains. Army and civil-service officers not at the top level would send their families to small mountain towns called hill stations, while they stayed behind to manage the native masses. The hill stations developed an unusual British countryside flavor, with churches and English bungalows on forested slopes. Social life in these towns was dominated by ladies and children, revolving around Victorian teas and balls. Young men on leave from their jobs in the heat or on temporary assignment to the hill station were in high demand at these parties to combat boredom, feed gossip, and satisfy forbidden passions. Many hill stations still retain an English flavor in their buildings, albeit now all inhabitants are dressed in *shalwar kameez.*

We land in Islamabad before dawn, and as soon as the airplane door opens, a cocktail of airplane exhaust and tropical heat overwhelms me. In seconds, sweat soaks my shirt, even though the sun has not cracked the horizon. I hope to only spend a few days in Islamabad, just enough to get the fossil-collecting gear and group together. My colleague Taseer Hussain is here already and has organized a red jeep and an old beige Land Rover for use as vehicles. With Taseer, everything is understated. He is a gentle man with an ironic smile and a Pakistani lilt.

"Well, Land Rover is old and may not come back, this is why you have jeep."

I like the Land Rover. It reminds me of the British nature shows I used to watch as a kid. I suppose it will go down fighting the harsh mountain terrain.

"Very good, who all is going to Skardu?"

"Two drivers, Munir and Raza, I believe you know them. Also, head cook Rookoon, assistant cook, and Mr. Arif."

"That seems like a lot of people."

"It is, everyone wants to see Skardu and cool off in mountains."

I think to myself that I want to be part of a fossil-collecting expedition, not a sightseeing tour, but I keep my mouth shut.

"You are not coming?"

"No, Karakorum Highway is only for young men like yourself. I will fly to Skardu after you arrive. You can pick me up at the airport there."

I consider the "only for young men," and wonder what he means, but decide not to ask. Moreover, I have no interest in flying there. The drive will be spectacular.

"I did not know that you could fly to Skardu."

"Oh, yes. Flight is very erratic, it only takes off when weather is good. Landing is very tricky there. Pilot has to circle down to land. Valley is too small for a straight approach."

OK, so now I am happy not to fly. Skardu is a small town in a valley on the Indus (figure 1). Much of that river runs in a narrow gorge in the mountains, but at Skardu the valley broadens and some agriculture is possible. Driving to Skardu will be one long amazing geology lesson, with the mountains as teachers. We will cut through several mountain chains covering about four hundred miles from south to north, and together called the Himalayas.[1] The different ranges have very different geological histories, but they are all associated with what may be the greatest geological event in the recent history of Earth: the collision of the Indian continent with Asia and the obliteration of the sea between them. In this

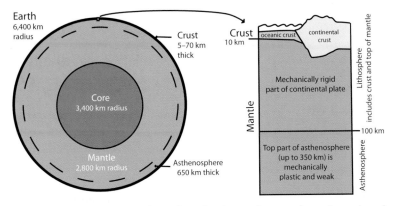

FIGURE 27. Cross-section of the earth, with a tiny section near the surface enlarged to show the different layers. All numbers rounded and approximate.

sea whales originated, and from its bottom the Himalayas rose. All along the way, the effects of this process will be on display. Even cooler, the process is still going on. The Himalayas are still rising.

If you could cut through a continent with a giant knife, you would see that the part that we walk on is just a thin shell, the crust of the earth (figure 27). There are two kinds of crust: continental crust, which makes up most of the land and underlies the shallow ocean near the coast, and oceanic crust, which forms most of the deeper ocean floor. The continental crust is between twenty-five and seventy kilometers thick, whereas oceanic crust is only five to ten kilometers. On the globe, there is much more oceanic crust than continental crust. The two types of crust behave differently and are part of large independent masses that move with regard to each other. Geologists call these plates, and the process of their movement is called plate tectonics. Imagine that the crust is like ice on a frozen-over swimming pool. When the ice breaks, slabs of it will move with respect to each other. When two slabs collide, one will go underneath the other, and the top one may rise out of the water. On Earth, if one of those pieces is continental crust and the other oceanic crust, the oceanic crust, being heavier, will go underneath the continental crust, a process called subduction. The subducted slab will slowly melt as it goes deeper underneath the crust. The molten rock, now lighter than its surroundings because it has expanded, will rise, break through the layers above it, and form rows of volcanoes all along the margin where the subduction is taking place. When two slabs of continental crust collide, neither goes down in an orderly fashion.

Instead, their edges fray, crumble, and crash on top of one another in a chaotic pattern. This is mountain formation.

The reason that the plates move at all is much deeper below the surface. Some one hundred kilometers underneath the surface of the Earth, there is a zone where rocks are in a semi-solid, semi-liquid state. That zone is continuous around the earth and is called the asthenosphere. The asthenosphere flows, and the plates with their continents and oceanic crust float on this layer. In our swimming-pool analogy, the slabs of ice move because the water that they are floating on actually flows.

The concept that the earth's crust is not constant but consists of plates that are movable with respect to each other was revolutionary, and led to a tidal wave of insight in geology in the 1960s. However, it all started with a German scientist, Alfred Wegener. Wegener was trained as an astronomer, but worked most of his life studying the weather. While in his university library in 1911, Wegener found a list of fossil animals and plants that occurred on both sides of the Atlantic Ocean. An important clue was *Mesosaurus,* a dog-sized reptile (not to be confused with the more famous and unrelated *Mosasaurus*). Fossils of *Mesosaurus* are only found on the western side of southern Africa and the eastern side of southern South America. *Mesosaurus* only lived in freshwater, and it was not clear how it could have crossed the Atlantic Ocean. Wegener searched for evidence from other fields of science, and he found it in geological structures. The Scottish Highlands were similar in structure to the Appalachian Mountains, for instance. He then found records of fossils in places where the present climate certainly would not support them—fern fossils from Spitsbergen, for instance. This kind of evidence led him to believe that the continents moved. He called his theory *continental drift*, and published it in 1915. Wegener was taken to task for his idea by other scientists. Rollin T. Chamberlin of the University of Chicago commented: "Wegener's hypothesis in general is of the footloose type, in that it takes considerable liberty with our globe, and is less bound by restrictions or tied down by awkward, ugly facts than most of its rival theories."[2]

Wegener's theory had its problems, especially that he did not know by which mechanism the plates moved. At that time, it seemed preposterous that such giant objects as continents could drift. But plate tectonics is now generally accepted by scientists and laypeople alike. I find Wegener's story interesting because it took somebody from a discipline outside of geology to get a great insight that tied together a large body of incoherent facts within geology. It seems that geologists at the time could not see the forest for the trees. It took an outsider to stand back and see the forest.

With India, it all started 140 million years ago, when dinosaurs were the dominant animals on Earth. Currents in the half-molten depths underneath Africa pulled on the solid rock above them, breaking apart the African plate by means of two giant cracks. The African plate split in three, from west to east: Africa, Madagascar, and India. The cracks grew, and the ocean flooded them. As the continents drifted apart, the growing cracks between them were filled in with new oceanic crust: molten rock moved up, solidifying when it reached water and making new ocean floor. These are the mid-ocean ridges.

The first rift, between Africa and Madagascar, was short-lived. It stopped growing and resulted in a narrow strait between Madagascar and Africa. The second rift, on the other hand, continued to open, and is still growing. The Indian plate is moving north on one side, away from Africa and Madagascar on the other.

Plate tectonics is on my mind as we drive north, away from Islamabad, toward the edge of the Indian Plate (figure 1). It gets cooler as the Karakorum Highway enters the front ranges of the high mountains. The Indus rages here, and rips into the mountains that tower along it. It is a different river from the sluggish, broad, mature one I know in the plains.

On the second day of our drive, we enter the Kohistan region of the Hindu Kush mountains, and the Indus canyon widens into a broad valley. It is a wild region. There have been kidnappings of foreign trekkers, and the Pakistanis in my group, all from Punjab Province, say Kohistani people cannot be trusted. The landscape is monotonous, colorless. The name of these mountains means "Hindu killer" and refers to the fact that, some generations ago, no non-Muslim could travel here and live. The mountains are broad and barren, brown and beige. I imagine them as the enormous shoulders and heads of an army of giants that has been buried here upright. Small villages become visible between the shoulders of the giants, in small side valleys, with buildings made of local rock, all brown and beige. Mountain streams crash from the heads of the giants into the Indus. They are brown too, eroding the giants' brow. In spite of those streams, the land is dry, lacking plants; beige dust covers buildings, men, and beasts, like an old faded postcard. Everything is in shades of brown. The red jeep is not red anymore. The Land Rover is even beiger than it was as it drives on the dust-caked asphalt.

Kohistan was an island before the India-Asia collision—it was positioned in the sea that separated the continents. As the collision took place, Kohistan was clamped by a vice made of the northern edge of India and the southern edge of Asia. Kohistan is large. We drive for the better part of a day to cross it. I can see far down the long and straight valley, but the mountains to the sides frame my view. My eyes get used to the hues. It seems peaceful and slow.

Suddenly, I sit up with a shock, blinking and staring. Down the valley, far away, where the brown mountains meet the horizon, another object appears. There is a new and massive mountain, not brown, but made from black rocks and topped by white snow, much farther away than the familiar Hindu Kush, but still easily towering over them. The color scheme is unsettling and discordant.

It is Nanga Parbat that asserts itself with majestic dominance: the ninth-highest mountain in the world, nearly twice as high as the mountains near it. It is rarely climbed. The weather is treacherous. Storms materialize very quickly, giving climbers no time to find shelter. The geology of Nanga Parbat is fascinating. This mountain is part of the Himalaya Mountains, not the Hindu Kush. The Himalayas, in their strict sense, are the mountains at the northern part of the Indian Plate, south of the Hindu Kush island. The continental collision started about fifty million years ago when the advancing Indian Plate captured the island blocks between it and Asia and sutured them into one landmass. In the process, the northern fringe of the Indian Plate was crushed too, making the Himalaya Mountains. All of these continental blocks, India, Asia, as well as the islands, had continental shelves—the shallow seas surrounding the land mass that are geologically more like continents than like oceans. The beginnings of the collision sutured continental shelves together, while a shallow sea still separated land masses. As the collision proceeded, it resembled less the mutual crumbling and crushing of two colliding cars, and more the collision of a large truck and a car, where much of the car was forced underneath the truck. India was the car, and about two thousand kilometers of its northern edge was forced down, underneath the truck, the Asian Plate. However, one stubborn part of the car refused to go down, and managed to override the truck. That part is Nanga Parbat. There it stands, different from its surroundings and proud of it. It is not often that I am humbled by what I see, but here, I am.

The geology relates to the fossils in a very direct way. The whales were living in and around the shallow continental shelf along the edge of Indian plate. The sea that they knew, the Tethys, would disappear within a few million years; but before that, they would go extinct, replaced by the whales that would conquer all of Earth's oceans.

Kohistan is a rough country, and it grates on the tempers of our crew. I ride in the red jeep. Its driver is Munir, a tall Sunni Muslim in his thirties. He stops in a small village. Behind us is the Land Rover. Its driver, Raza, is older, smaller, and a Shiite. He stops, too. He has the frame of a street fighter. He runs to Munir and starts shouting. Munir screams back, there is pushing, Raza punches, Munir ducks and punches back. Little Mr. Arif, much older and smaller and a very thin man, jumps between the fighters. They drop their fists. Munir runs to his car shouting the word *bohti* over and over again. His passengers, Rookoon and I, also run, not wanting to be left behind by Munir. We speed away, billowing brown dust, much too fast to avoid potholes. Munir speaks angrily to Rookoon but eventually calms down in what appears to be desperation, even though I cannot understand a word he says. I do not get the story till we stop, hours later, when Arif explains that the village where we stopped is known in the rest of Pakistan for biting black flies that carry disease. Munir was hungry—*bhoti* is Punjabi for "meat chunk"—but Raza was afraid of getting bitten and getting sick. Kohistan pushed their muted dislike for one another into a fistfight.

Nanga Parbat is now close and looms over us. We are nearing the place where it slid over the rocks it conquered. Geologically, there is chaos here. Blocks the size of houses and composed of all different rock types have been strewn around as the battle raged between the giants over who would be subducted and who uplifted. The fight continues, and blocks tumble down into the Indus Valley as Nanga Parbat reaches higher into the sky. The brown Indus fumes angrily as it seeks passage around obstacles. It reminds me of a van Gogh painting from late in his life: too wild to be real, broad brush strokes that shout for attention out of tune, drowning out the big pattern unless you step way back and squint your eyes.

Nanga Parbat has pushed the Indus off course. The river was running east-west, but the mountain pushed it to the north, so now it flows around the mountain on three sides before continuing west. We follow the river upstream, leaving the Karakorum Highway. The river now is in a narrower valley that will lead us to Skardu. We enter a third set of mountains, the Karakorum Range, originally part of the Asian Plate, with high, sharp peaks, and the home of K2, the second-highest mountain in the world. Forced into its narrow valley, the Indus is now incessantly furious. The villages are tiny, perched on small fans of rock rubble tumbling out of small side valleys. Their houses are built on top of each other—one person's roof is another person's floor, slightly offset, like steps of stairs.

The villages on the south wall of the valley are always in the shadow of their cliff. Each village has it is own hewn-out terraces, with narrow, steep paths connecting them. We stop and are surrounded by kids. Just as the terrain is a blend of geological terrains, so are the kids a blend of races. There are some with the coffee-colored skin of South Asia, some with Chinese features, and paler ones with almond and green eyes, like Afghans. They're all Pakistanis, but show traits of the conquerors and rulers of these lands, Mongols from the north, Afghans from the west, Sikhs and Moghuls from the south. This road is the only way over land to get to Skardu and its associated villages, so there are many trucks. Our rest site is a rest stop for them too. Men from the village walk around soliciting oil changes. If the driver consents, they crawl underneath the engine, and open a valve; black oil runs onto the sand. The valve is closed, and new oil added. The used oil stands in puddles in the sand, dozens of shiny black lakes, eventually adding a glistening sheen to the Indus raging below. The lack of environmental conscience depresses me.

The valley narrows, the road climbs and falls, and when opposing traffic comes, we have to pull off to let them pass, slamming on the brakes if someone cuts a corner around a promontory. The road goes down, seemingly directed straight into the furious Indus and its deafening concert of raging waters. From my position in the car, I cannot see sky. The mountains are too high, the valley too narrow. It is dark around me. We are near the bottom of the valley. I feel panic: this is how I imagine the River Styx opening into the underworld.

Skardu is in a pleasant valley, and as close to the end of the world as one may go. One road goes north from it, ending at the foot of K2. Another road goes east to the closed border with India. The location of the border is disputed, so the practical edge of Pakistan is the Line of Control, a cease-fire line from one of its wars with India, monitored by United Nations observers.

The fieldwork is mostly a bust. Roads are eroded away. It is impossible to cover much terrain on day trips, and hiking to locations seen on a map requires mountaineering skills, which I lack. The army stops us and sends us back from sensitive areas. In spite of that, I learn about the mountains and their geology, and enjoy the scenery and the people. As far as fossils go, this field season will have to be carried by a few days that we will spend down in the Kala Chitta Hills, down on the hot Indus Plain. We want to revisit the *Ambulocetus* locality and dig deeper, getting the rest of our prized skeleton, if there is anything still buried. As we drive the Karakorum Highway back toward the plains, the high mountains reinforce the lessons they taught me about plate tectonics. I also think about the first geological explorers of this area, people who provided the foundations for the work I am doing here, but who had no idea about plate tectonics.

KIDNAPPING IN THE HILLS

Long before plate tectonics was a generally accepted way to think about the world, the first South Asian fossils relevant to the origin of whales were collected by T.G.B. Davies in the Kala Chittas. At the time, the area was part of British India and Davies was a geological surveyor for the Attock Oil Company. He was sent there to investigate reports of oil seeps, places where oil was exuded by rocks. In 1935, Davies drew a geological map to determine whether oil exploration was feasible. While mapping in the field, Davies also collected some fossils, and these eventually ended up on the desk of a vertebrate paleontologist at the British Museum of Natural History in London, Guy Pilgrim, who had been the paleontologist for the Geological Survey of India.[3] In 1938, with World War II looming, Richard Dehm, a professor from Munich, Germany, visited Pilgrim to see his collection of Indian Miocene fossils (thirty-five million years and younger), with the intention of setting off to British India and collecting some himself.

I visited Dehm in the mid-1990s, when he lived in a retirement home in Munich. Dehm was excited to tell his story, knowing that I was now working in the area where he collected half a century earlier. On his visit to London, Pilgrim had also shown him Davies's collection of Eocene fossils from the Kala Chitta Hills. Pilgrim encouraged Dehm to go there too. Dehm set off on a long journey in 1939, sailing past the Cape of Good Hope, collecting in many places in India, and then going on to Australia. He was in Australia when the war caught up with him. Being German, he was jailed, but eventually he was released and he traveled back to Germany. His fossils were confiscated by the French, and remained on the ship he had been traveling with, moored in a French port, with the frontlines of the war moving across France. Eventually, the Germans conquered France's west coast, and found the ship, and Dehm was reunited with his fossils. Dehm was not a Nazi. As a matter of fact, the Nazis disliked him and moved him from his important museum position in Munich, the Bavarian and Nazi heartland, to the small provincial outpost of Strasbourg, freshly conquered from the French. There he spent the duration of the war.

Dehm returned to Munich after the war and made plans to go back to British India to collect more fossils. The war had not only left Europe scarred; British India was broken up into Hindu-dominated India and Muslim Pakistan in 1947. The Kala Chitta Hills were now in Pakistan, and Dehm visited his old sites in this young country in 1955. His collection grew, and he published it in 1958.[4] Dehm had found a small jaw fragment with two teeth. It belonged to a whale; however, he was not aware of this. No whales older than *Basilosaurus* were known, and Dehm's whale was too different and too fragmentary for recognition. He did make a remarkable inference, though, guessing a diet that befitted a cetacean for the animal. He called the animal *Ichthyolestes*; *ichthus* means fish in Greek, and *lestes* means robber. It was the first Indo-Pakistani whale to be named, and remains one of the oldest whales in the world. The specimen came from rocks a few hundred yards from the ones that yielded a jaw which Robert West, thirty years later, identified as a whale.[5] Dehm marked his site on a hand-colored copy of the map that Davies had made. He gave me his map when he heard I was working there.

I think of Dehm and West as we drive toward Attock. In 1987, I flew to a conference of the Society of Vertebrate Paleontology in Tucson, Arizona. On the plane, I happened to sit next to West. I told him that I was interested in working in the Eocene of Pakistan, and asked him whether he minded if I were to visit his old sites. There is an unwritten rule, observed by many paleontologists but also frequently broken, that one does not visit localities where someone else is working without their permission. West had not worked those sites for years, but I wanted to make sure. Graciously, he said I should go ahead and that he had no claims to those areas.

We leave the Himalayas and enter the frying pan that the plains of Punjab are. Temperatures are in the low 100s, and the humidity is high. We stay at a railroad guesthouse in Attock, next to the tracks and the station. The town is scorched, dusty, and miserable; the smell of rotting garbage and diesel fuel fills the air. But the guesthouse has a courtyard that smells sweet, it is full of vivid colors, there are flowers everywhere. A caretaker is employed at the guesthouse full-time, and his main job is watering this visual paradise. But it does not diminish the heat. I wear sunglasses and a hat, but it still gives me a headache, and the sweating dehydrates me. The electric service is out. Arif and I share a room, and the first thing he does is position his bed in the dead center of the room. Odd, it appears to me, as I leave my bed along a wall. Later, when the electricity revives, the ceiling fan comes to life. It is located right over Arif's bed.

The next morning we leave the guesthouse at four A.M., hoping to arrive in the field area at sunrise and leave before the hottest part of the day. Attock is dark and still asleep, no breakfast available, we see no person or beast. Munir drives the red jeep. We are tired. No good sleep is possible in this heat. We cross the high parts of the Kala Chitta Hills, still not having seen a soul. The hills here are the tiny cousins of the mountains that we left a few days ago; they are the last ripples of the continental upset to the north. There are gray Jurassic limestones, formed in oceans that harbored ichthyosaurs and plesiosaurs, long before whales originated. A stretch of road crosses an area covered in bushes and labeled "dense jungle" on the map. *Ambulocetus* is just a mile away. The bushes are taller than a man and grow in clusters. You can easily walk around them, but you can never see far. The place is like a giant, green maze. We round a hairpin turn, and suddenly we are all fully awake and in shock. The place swarms with police in army-like outfits. There are busses that have brought ground troops with semi-automatic guns, camouflaged jeeps, and armored vehicles. They stop us, and ask what we want. Arif explains in a nervous voice. They tell us to keep driving, we are not to stop in these hills but to cross the hills to the police station in Basal, south of the hills. A man has been kidnapped, and the kidnappers are hiding in this jungle. Today, the police will hunt them down. I am puzzled and dazed and try to organize my thoughts on the way to Basal. They can't do this to us. I have waited for two years to come back here. We cannot just give up. The police in Basal *have* to give us access. I am gearing up to make my case.

The police station in Basal is built around a courtyard. We park the car outside. Arif alone passes the guards. He does not want to take me inside. I think that he is worried that the presence of a foreigner complicates matters further. He comes back with no news at all. I barrage him with questions, but he remains silent. Did he make a case at all, or did he just go inside because I insisted? I call Taseer, who is in a hotel in Islamabad. He says to return to Islamabad. The next day, an Islamabad newspaper reports that four policemen have been killed in the operation and that the kidnappers were not caught. Taseer tells me to go home and forget about it, the fossils will be safe in the ground for another year. I am disappointed.

Life goes on. The parts of *Ambulocetus* that are still under the ground remain where they have been for forty-eight million years.[6] But the kidnapping incident is part of a larger problem with Pakistan. Too often, when I point at a place on the map, Arif tells me that the place is off limits for security reasons. Pakistan is too risky a place to be the sole purveyor of study material. So, I am looking elsewhere. There are fossil whales in India. In addition, India is opening up politically.

INDIAN WHALES

The man who looms larger than life in Indian paleontology is Ashok Sahni, the father of vertebrate paleontology in that country. Some years ago, I sent him a letter asking him about his study of the tooth enamel of Eocene Indian whales.[7] A letter came back saying little about enamel and instead inviting me to visit his lab. It was a pleasant surprise.

Studying tooth enamel in whales could be very interesting because it may provide clues to the strange tooth wear in ambulocetids and basilosaurids. And that may help us understand what these animals were eating. Of course, to study enamel, one has to cut a tooth on a diamond-blade saw and look at the cut face with an electron microscope.[8] That destroys the precious specimen. Having more fossils would make it hurt less to cut some of them up. Teeth from Indian whales would be a welcome addition to those from Pakistan. I am taking Dr. Sahni up on his invitation.

CHAPTER 6

Passage to India

STRANDED IN DELHI

Islamabad, Pakistan, February 1992. After a month of fieldwork in Pakistan, I am boarding a plane that will take me to India's capital, New Delhi. Flights between the countries only happen twice a week, the result of the hostile stance between them. Soldiers commonly shoot at each other across the Line of Control, near Skardu. I am excited to go to this new country, meet Ashok Sahni and his colleagues, and study their whale collections from Gujarat, the western Indian state on the Indian Ocean. The original arrangements were all made by airmail back and forth, weeks between correspondences. From Pakistan, I tried to call Ashok but never reached him, so I hope for the best. The Air India flight to India is crammed with Indian Muslims who have visited Pakistani relatives. As we land, it is prayer time, and many of them unroll their prayer mats in the corridors of the airport to pray, obstructing the flow of traffic. Airport officials, Hindus and Sikhs mostly, in drab military-looking uniforms, give them a condescending look, but let them be. Coming down the escalator in Delhi, I see a large wooden statue of Ganesha, the elephant-headed Hindu god of travelers and traders, welcoming all those who want to be welcomed. Coming from Muslim Pakistan, where depicting deities is sacrilege, I am taken aback by such blatant idol worship, but also elated to enter this strange new world. Back in the United States, I was unable to buy a ticket to fly from Delhi to Chandigarh, so I will buy it here, at the airport, for the flight tomorrow morning. I ask at the Air India check-in desk to buy a ticket to Chandigarh.

"We do not fly to Chandigarh."

I am puzzled. I thought they did. I don't really believe her. "So, where can I buy a ticket?"

"You go outside."

I walk on, leaving the terminal building. Then I realize my mistake: the flight to Chandigarh is on Indian Airlines, not Air India. I turn around and walk back into the terminal building. A man stops me. "Change money, sir?" There is a rich black market in money-changing here, and I politely decline the offer.

A policeman stops me and asks for my ticket.

"I do not have one yet, I will buy it inside."

"No entry without valid ticket."

"How can I buy one, when the ticket counter is inside? I have to go inside."

"No entry without valid ticket."

From his face, I can see that he means business, so I turn around, somewhat panicked.

"Hotel, sir?" A man is actually grabbing my bag already. He pulls on it. I jerk it back hard, and growl, "No."

I ask another man where to buy tickets for Indian Airlines. He points me toward a small office outside the terminal. On the way, four different men insist on offering taxis, hotels, and money-changing. When I reach the office, it is closed. Why is that? It is only three P.M.

"Taxi, sir?" A turbaned Sikh looks hopeful. I shake my head, but he is not convinced. "Very good, sir, you need hotel, I will take you there, best quality." The thought of being pushed into a taxi by a hawker and being passively taken anywhere in Delhi is not comforting.

"No."

I walk up to another man. "Where can I buy a ticket for Indian Airlines?"

He points at the office from which I just came.

"It is closed," I say.

"Yes, is closed," he says, and walks on.

I am puzzled. I don't know how to proceed. A policeman walks up to me. "Do you want to change money?"

This is scary. If I say yes, he could arrest me for changing on the black market. If I say no, he could arrest me for something else, just because he is upset with me. I decline, but with fear in my eyes.

All the local lowlifes have now figured out that I don't know what I am doing, and they all have great ideas about what I could be doing.

"Sir, taxi?"

"You come here."

"Hotel, sir, very nice, please come."

My nerves are fraying. Everything is going wrong, and I cannot see a way to turn it around. I throw in the towel on the ticket and decide to work only on saving life and sanity. In order to get rid of the masses, I need to appear resolute, so I walk up to a queue of people standing in line and join them. I do not know what they are in line for, but it must be something legitimate and possible, and I hope that it implies to the masses that I have made up my mind and am doing something that does not involve them. It buys time. My nerves settle a bit. I can think without being accosted. Best to go to a hotel and see if the desk there can help me with the flight. But which hotel? I need help from someone who is not after my money. The man in front of me is well dressed and is fully ignoring me—a good sign.

"Sir, can you tell me the name of a good hotel nearby?"

He checks me out and smiles, no doubt noticing the stressed look on my face. "You can go to Ashoka Palace, very good."

The words sound like music—Ashoka Palace, and "you can go" as opposed to "please go here." It allows me initiative and gives me courage.

"Thank you very much. Can you tell me at what time the office of Indian Airlines will reopen?"

"It will not open."

"Why not?"

"It will open tomorrow. Today is Republic Day."

That clears up that issue. Republic Day is a national holiday, similar to Independence Day. No wonder everything is closed. To the Indians I asked, it was obvious that the office would be closed today and that my question was for future reference. They were not trying to mislead me at all. So actually, there is nothing left to do but go to a hotel and wait for the morning.

I walk to a taxi man who was not involved in mobbing me earlier. After a short skirmish over the fare (there are no working meters in Delhi), he takes me to the "Palace." I take in stride the driver's habit of letting go of the wheel and assuming a praying posture with his hands whenever we drive by a Hindu temple, and we reach Ashoka Palace without further complications. I secretly celebrate living through the forty-five most mentally harrowing minutes of my life, having interacted with thirty-six different strangers (indeed, I counted them), nearly all of it tense and confusing. In the room, I drop my backpack on the floor, lie down in bed, and immediately fall asleep—to be awoken, a few minutes later, by the man at the reception desk, who asks if I want to change money at a good rate. I rudely tell him that I do not and slam down the phone.

The next day I go back to the airport and buy my ticket without further incident. I fly to Chandigarh, the capital of the state of Panjab, once united with the differently spelled Pakistani state of Punjab, before the countries were divided.[1] Sahni picks me up at the airport, and we drive to Panjab University. He laughs at my experiences in Delhi and says that traveling by train is much to be preferred over Indian Airlines. The drive delights me. If you ignore the living beings, Chandigarh is a modern city, with straight boulevards, whitewashed houses built in straight rows, and many traffic circles. However, this attempt at organization is altogether insufficient to subdue India's ruling Muse of Chaos, and like all other Indian towns, cows, dogs, boys with kites, and girls guarding littler siblings dominate the street; tea vendors set up shops in wooden crates, fruit salesmen stack their wares high on rickety pushcarts, fully blocking one traffic lane, and grown men just hang out as if they were teenagers at the mall, smoking and chatting, watching the world with the uncomprehending disengagement that one has watching an anthill.

Chandigarh is the capital of Panjab, the homeland of the Sikhs, some of the more striking inhabitants of this subcontinent. The men sport fully bearded faces and colorful, tightly rolled turbans that hold the long hair which, following the rules of their religion, should not be cut, ever. There is a minor Sikh rebellion going on. Some Sikhs are trying to gain independence from India. The Indian government is cracking down hard on them, with mixed success. There are sandbagged holdouts on campus, filled with men in army fatigues brandishing machine guns. A Red Cross sign near campus proclaims: "Don't shed blood, donate it." Walking onto campus, I run into a demonstration, people with signs I cannot read and angry slogans, but my worries dissipate when Sahni explains that they are just employees demanding higher pay. Rebellion or not, raises are important too.

Sahni introduces his students, among them Sunil Bajpai, a quiet fellow of my age, mustached and already balding. The whale fossils are about five to ten million years younger than the Pakistani ones, and they are the subject of Sunil's thesis. Fossils were first discovered here during geological survey work[2] by the Geological Survey of India, when India was still a British colony, but Sahni's lab was the first to start looking for fossil whales seriously.[3] The most important publication is Sahni and Mishra's 1975 monograph, of which there are only few copies in the United States.[4] I now see it for the first time in the original and it feels rather like a pirate's treasure map as I open it in the dimly lit, cavernous lab—pages nearly two feet high, printed on musty yellow-brown paper that is frayed at the edges and covered with dirty thumb prints and mildew stains. There are photos of fossils, but they are all out of focus and printed with too much contrast. The fossils are like white islands in a black ocean, the dashed lines indicating anatomical features like secret trails that lead to buried treasure. I am glad I came here. The fossils might as well be pirate's gold.

Sahni and Mishra were remarkable. Mishra, with not even a car, collected many fossils and described the geology, and together they recognized that these fossils represented early whales, well before the Pakistani fossils were recognized as such. Truly, these whales do not look like modern whales, and the fossils were mere fragments. It was an accomplishment to identify them as whales, more so since the famous Smithsonian paleontologist Remington Kellogg had written that the early whales "evidently did not reach the Indian Ocean during Eocene time."[5]

Sahni also shows me the fossils, most of them wrapped in yellowed newsprint and hidden behind fossil elephant tusks and antelope skulls. The Indian whales now come to life for me. There is a blackened skull, with a long snout and eyes the size of marbles. Its ears are large and far apart, different from *Ambulocetus* and *Pakicetus*. Sahni and Mishra named this whale *Remingtonocetus,* after their Smithsonian colleague.

Sunil and I make tentative plans to work together in the place where these whales come from, an area called Kutch near the Indian Ocean in the state of Gujarat. Combining the Indian whales with *Pakicetus* and *Ambulocetus* will allow me to study three snapshots of the rapidly evolving whales just a few million years apart—a unique opportunity.

Kutch is about six hundred miles from Chandigarh, too far for a quick look around on this there-and-back trip to India. However, there are some localities with Eocene whales within driving distance in the Indian high Himalayas, and another whale, the ambulocetid *Himalayacetus,* was found there. So, the next day, Sahni takes me and his three students into the mountains, a three-hour drive in Sahni's tiny van.

The fossil sites in the Himalayas are different from those in Pakistan. It rains a lot here. Beautiful and peaceful pine forests cover the slopes, dampening sound and light, but bad for collecting fossils. Rocks are covered with pine needles or undergrowth. The only outcrops, places where bare rocks and fossils are visible, occur in the small streams. But the exposures are steep and slippery, and weeds compete with my feet for a foothold. The fossils here are found in shale—compacted mud, really. Shale is friable and unstable; the surface is slippery and full of loose rock. I spend at least as much time seeing where to put my feet as I do looking for fossils. To make matters worse, it is starting to rain, and the people in this village use this valley as the community bathroom. Sahni and I have raincoats, but the three students are miserable in their wool sweaters. Sahni shows no mercy and tells them they should have brought coats. One of the students is from this area, and he's used to scrambling on these slopes. He finds part of a tooth. It is a whale tooth, and I notice that it has a clear break on one side. The remainder of the tooth is still in the rock. I did not bring tools to extract it, or glue to join the pieces, not thinking fossils would be found. With my pocket knife, I flick away the shale that surrounds the tooth. It comes out in one piece, and I wrap both pieces in my handkerchief and put them in my shirt pocket. Amazing—a whale tooth collected from a very difficult place under very poor conditions.

The student asks us to visit his village and his house. Sahni agrees reluctantly. It would be somewhat of an insult to say no, and an honor for the family to receive a professor from Chandigarh. The houses are built along a steep hill, and, just as in the Pakistani mountains, the roofs of one family are the floor of their neighbors above them. The houses are mostly wood, weathered gray and unpainted; glass windows are small, and wood lattice covers them, very picturesque. We scramble up the steep path and drop in, unannounced, on his family. His mother and sister are home, and offer some food. Again Sahni hesitates. This is a poor village. He does not want to impose. She brings out a dish, which is devoured quickly by the three young men, their professor, and his foreign guest. Another dish comes out, also eaten; another; and another. Dishes keep coming. A large meal had been prepared for someone, and now it is all gone. Sahni is supremely embarrassed. His group wolfed down a feast obviously not intended for them. But there was no way out. Refusing to eat would have been very rude, in this culture that honors guests to an extreme.

On the way back to the car, walking to the lower parts of the village, I notice a stray dog eating something. I take a professional interest in road kills, and often collect them, since the bones are useful to compare to fossils. I chase the dog away and walk over to the morsel he dropped.

My heart stops as I see what it is. *The dog was eating a cut-off hand.* The shock only lasts a second, as I realize that the hand is hairy and much narrower than a human hand. It is a hand of a monkey. My heart still pounds, and I quickly walk away. Later, back on the road, we see macaque monkeys sitting on the side of the road. The rain worsens as we descend from the Himalayas through numerous hairpin turns. The windshield is opaque with rain, and Sahni has the inexplicable habit of turning on the wipers only for one or two beats, only after all visibility is lost, and then turning them off again. I bite my lip and sit in the passenger's seat, but all ends well. Traffic is the most dangerous thing when visiting India.

Back in Chandigarh, I use a sewing needle to clean the tooth. Sahni offers some Quickfix, an Indian version of model-aircraft glue. The slogan on the box says it "joins everything except broken hearts." The glue's wispy threads fly uncontrollably; it does not adhere very well at all. I resolve to bring my own glue next time. To make matters worse, the only little brush available for gluing has metallic silver paint on it. The tooth I put together has a silvery sheen, a permanent reminder of this memorable trip.[6]

WHALES IN THE DESERT

Following up on that first visit to India, Sunil and I make plans to collect in Kutch in 1996. Having learned from experience, I now fly directly from the Unites States to India and meet him in Gujarat. In the past, Sunil has come here often, working on a shoestring. In the morning, he would take the bus to a village in the desert, talking the driver into dropping him off at localities on the bus line. Busses are the main mode of transportation for villagers. People are too poor to have personal cars; besides busses and trucks, there is rarely a car on these roads. The bus reaches villages only twice a day, and the schedule does not take the needs of the fossil collector into account. As a result, most of Sunil's localities are near bus lines. He would often arrive at a site before the sun was up, wait for it to get light, and then leave when the bus came by on its return journey in the early afternoon. It gave him little time to collect, and he was limited in the amount of tools he could bring and the number of fossils he could collect. The collection he put together is surprisingly large.

We now have a small grant from the National Geographic Society and are able to rent a car. In India, one cannot rent just a car, one always rents it with driver. Naveem is a quiet and gentle man and, importantly to me, a careful driver. The car is an Ambassador, built in India, without four-wheel drive. Naveem takes pride in driving it off the paved road through streams and hogbacks. Only once do these get the better of him and we end up suspended on the bumpers while trying to cross a ditch. With a reluctant smile, he tells us that he will not use that track again.

Naveem drops us at Babia Hill, a flat-topped hill with intense weathering on its slopes. The nalas (dry streambeds) are all bone-dry early in the year; they only carry water in the monsoon season in the fall. Sunil finds some ribs weathering out of the wall of such a nala. They are lined up in a row and are as thick as broomsticks. A large thorn bush casts its irregular shadow over them. The fat ribs immediately identify the fossil as sirenian, and I notice that another row of ribs is on the other side of the thorn bush, also nicely lined up. This was a chest, buried in its entirety. I can't wait to see the vertebrae between the ribs, underneath the thorn bush. I hack at the bush with my knife, and the excavation starts. The vertebrae show that the head side of the beast pointed into the nala, so the skull was probably washed downstream during some monsoon flooding. It could have been last year, or a thousand years ago. On the off chance that it was recently, I leave my backpack with the sirenian chest and walk down the nala. A short walk down, a piece of bone sticks out of the bank. I dig into the bank with the back of my hammer. It is a jaw—I see the cavities for teeth, and, deeper, a tooth! However, it is not a sirenian tooth but a beautiful whale tooth. I expose a second tooth with my pocket knife. It is slow going, because the more appropriate tools are in my backpack. The jaw goes deeper still, into the wall. To get this out I should get my preparation tools—but I should really first finish the sirenian chest, and my quest for its head. As I struggle with priorities, Sunil comes and tells me that he has found something else. I leave my hammer with the jaw as a marker, so I can easily find its spot in the twisting nala, and follow Sunil. Another short walk, and he points at a white object protruding from a crack in the nala's side. The only tool I have now is my pocket knife. I try to cut away at the red brown mud. The fossil is bright white, a color that indicates gypsum here. Gypsum is a mineral that is dissolved in the water, and when it is abundant, it often replaces, molecule by molecule, fossil bones and teeth. Environments where this happens are often enclosed bays that are filled with seawater and that dry up. As the water evaporates, the gypsum concentration goes up, and eventually the gypsum forms crystals that replace the bones and teeth of an animal. Some gypsified fossils keep the shape of the original fossil quite well, but some deform it beyond recognition, and gypsum is hated by many paleontologists. I use my pocket knife to cut away the sediment around the fossil, and what is revealed looks good. It is a very narrow bone, with two rows of small circular holes lined up along its length, like the holes in a flute. Another whale jaw—the holes are the sockets for teeth!

Overwhelmed, I sit down on the dirt, resting my head in my hands, not knowing how to proceed. My tools and backpack are at one excavation, my hammer at another, and me and my knife at a third. Getting the two jaws out will take about an hour each, and the chest will take half a day. Too many fossils is a nice problem to have, but it remains a problem.

Eventually, I remove the fossils in the opposite order in which they were found. Neither fossil is similar to the whales I know so well from Pakistan. The flute look-alike belongs to a relative of the skull in Sahni's lab, a remingtonocetid (to be discussed in chapter 8), and the other jaw is from a protocetid whale (to be discussed in chapter 12).

If *Ambulocetus* bridges the gap between land mammals and whales, these Indian fossils could bridge the gap between the Pakistani fossils and basilosaurids. If whale origins were a puzzle, we previously just had some intriguing pieces, but could not make out the image on the puzzle. With the addition of the Indian fossils, we can find enough pieces so that the image may become clear—as long as we have the time and money to collect them.

A 150-POUND SKULL

We are back in Kutch a year later, searching old and new localities. At Rato Nala, the road climbs a shallow escarpment that includes the outcrops of the Harudi Formation, a belt of rocks, three hundred yards wide, perpendicular to the road. It was a stretch of coast forty-two million years ago. Drab comes in many shades here: greenish, brownish, yellowish, mostly coloring muds. If you look closely, you see some brighter veins: thin layers of bright-yellow sulfurous rocks, glassy white gypsum layers, and nearly black seams of coal, never more than finger thick. The coal indicates abundant plant growth, marshes at the water's edge. To the eye, the most dominant type of rock is the Chocolate Limestone, a brown layer of limestone chock-full of bright-white snails and clams that indicate that this was ocean floor.

Looking north, beyond the Harudi, the place looks like a barren moonscape: miles of brick-red, blood-red, and black sandstones and

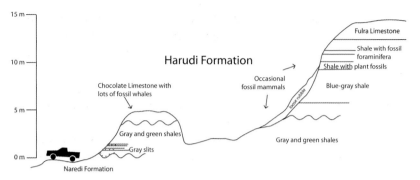

FIGURE 28. A geological section of the Eocene rocks of Kutch in western India.

mudstones, eroded into irregular shapes that make it hard to walk on (figure 28). This is the Naredi Formation, which was formed in a period of intense Eocene weathering. Rains leached the soil, leaving only the least soluble minerals behind. The dark colors are mostly iron oxides, rust basically. With all the nutrients gone, modern plants, too, find it difficult to live there. The Naredi was formed before the Harudi Formation. Looking south, the Harudi ends at the top of the escarpment. There, the Fulra Formation appears as a raised plateau consisting of blocks of hard and bright-yellow limestone with lots of clams and sea urchins, and snails the size of footballs, but no vertebrates. The Naredi rocks were formed when the land was exposed by weathering. After that, the ocean level rose, and the Harudi formed at a time when the coastline was here: muddy beaches, oyster banks, coastal swamps, and islands. The ocean kept rising and flooded more land, and there was a shallow, warm, and very productive sea, as documented by the reefs of the Fulra Formation. Geologists read the rocks as if they are the book that describes the history of a place. In Kutch, the book describes how, as the ocean flooded the land, a continental edge was slowly drowned.

The Chocolate Limestone forms the top of a row of low plateaus, flat-topped hills that can be up to sixty feet in height and half a mile in length. This is a good place to find whale fossils. We spend much of our time walking along the edges of the plateaus, where fossils become visible as they erode. A piece of bone on the slope triggers an intense on-your-knees investigation, head to the ground, scouring the surface. Even though this area is rich, a good day means having three fossils in your backpack upon return. Rich in fossils is a relative concept.

I like this place, partly because it is so remote. If you walk away from the road, you cannot see anything human-made, even though you can see for miles. The quiet is also beautiful. In the heat of the day, you can listen intently and hear absolutely nothing for minutes, when the quiet is just subtly disrupted by the faint hum of a distant insect flying by or a rare whisper of wind. The atmosphere makes me imagine looking back in time, when whales swam here.

Suddenly, Sunil wakes me from my musings as he calls from afar. He is running toward me, reaching me, exhausted, red-faced, catching his breath.

"Hans, Hans, I found a skull, the best skull I ever found."

We rush to the place. The specimen is totally embedded in Chocolate Limestone. We can see only the top of the skull, a ridge of bone that is the crest on top of a skull, the sagittal crest. It is embedded in limestone; the bone undulates in a pattern identical on the left and right side of the skull, wider where the eyes are, and forward onto the snout for about three feet. It feels as if I am standing on a boat on the ocean, and an Eocene whale is surfacing immediately next to me, only the top of his head emerging from the water. Sunil is right. This will be an amazing skull. The part that is popping out of the limestone is perfect. Our hammers loosen the baked dirt and my whisk broom sweeps it away. The limestone under our feet is not a massive layer; instead, it is broken up into large blocks. The piece with the whale skull is bigger than the others. It is the color of milk chocolate, with white snails and clams as marshmallows. But this is better than chocolate.

We excavate around it, and after a few hours, it is clear that the block is much too heavy to be lifted by one person and carried to the road, two miles off. We consider breaking it into pieces for transport. But the limestone breaks irregularly, and the incessant hammering causes cracks to form in places where they may easily pulverize fossil bone during transport. No, it has to come out as one piece. Dr. B. N. Tiwari, the third member of our group, solves the transportation problem. He cuts down two small trees, and we suspend the block from them with ropes. The driver takes our car on a circuitous route, finding flat spots, cross-country, toward us. There are just two hundred feet from fossil to car now. We suspend the fossil in its hammock. The slope is steep, slippery, and rock-littered. Like drunken sailors, the four of us stagger down, our load swaying with every step someone takes. Lifting it into the back of the car is not easy either, and the car sags perilously under its weight. But we make it to town. A local carpenter adapts a salvaged box to hold

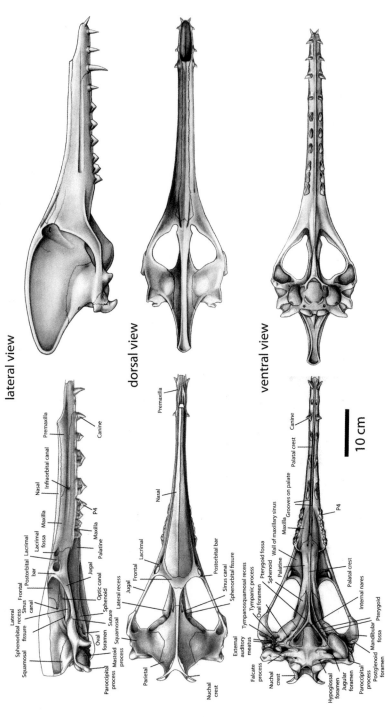

Andrewsiphius sloani

lateral view

dorsal view

ventral view

Remingtonocetus harudiensis

10 cm

Premaxilla
Canine
Nasal
Infraorbital canal
Maxilla
Lacrimal
fossa
Lacrimal
Postorbital
bar
Maxilla
P4
Palatine
Jugal
Frontal
Optic canal
Sphenoid
Sinus
canal
Lateral recess
Suture
Squamosal
Sphenotibial
fissure
Sphenoid
Lateral
recess
Sinus
fissure
Squamosal
Oval
foramen
Mastoid
process
Paroccipital
process

Premaxilla
Nasal
Lacrimal
Frontal
Jugal
Parietal
Postorbital bar
Sinus canal
Sphenorbital fissure
Nuchal
crest

Canine
Palatal crest
Grooves on palate
P4
Maxilla
Wall of maxillary sinus
Sphenoid
Palatine
Pterygoid fossa
Oval foramen
Tympanic process
Tympanosquamosal recess
External
auditory
meatus
Falcate
process
Nuchal
crest
Hypoglossal
foramen
Jugular
foramen
Paroccipital
process
Postglenoid
foramen
Mandibular
fossa
Pterygoid
Internal nares
Palatal crest

FIGURE 29. Skull of the fossil whales *Remingtonocetus harudiensis* and *Andrewsiphius sloani*, seen from three different angles. Drawing of *Remingtonocetus* is based on a single fossil (IITR-SB 2770, from S. Bajpai, S., J. G. M. Thewissen, and R.W. Conley, "Cranial Anatomy of Middle Eocene *Remingtonocetus* (Cetacea, Mammalia)," *Journal of Paleontology* 85 (2011): 703–18). *Andrewsiphius* is based on four fossils (IITR-SB 2517, 2724, 2907, and 3153, from J. G. M. Thewissen and S. Bajpai, "New Skeletal Material of *Andrewsiphius* and *Kutchicetus*, Two Eocene Cetaceans from India," *Journal of Paleontology* 83 (2009): 635–63). Both used with permission of the Paleontological Society.

the fossil. We stuff empty, tightly closed mineral water bottles around the block as shock absorbers. I am pleased with the improvised result as I prepare the specimen for shipment to the United States.

I am not pleased with the bill. Shipping it will cost over a thousand dollars. I do not have that much, so I leave the specimen with Sunil. The fossil finally reaches the United States some years later, and my fossil preparator spends a full year to extract the fossil from the block, knocking tiny pieces of rock off the fossil with a pen-sized jackhammer. The result is amazing. This is certainly the most beautiful whale skull that Kutch has ever produced (figure 29, skull on left).

CHAPTER 7

A Trip to the Beach

THE OUTER BANKS

Driving to the South Carolina Coast, 2002. I think of the long-extinct Indian whales as I drive with my family on a vacation trip to Kiawah Island in South Carolina. Weedy forests cover the mainland, like the "dense jungle" of the Pakistani maps, and they suddenly give way to flat marshes, swamps, and winding river channels at the shore. The bridge is long, but as we cross it, I can see the ocean across the island.

Geologists call islands like Kiawah barrier islands. They are basically sandbars that rise above the sea and grow when they are fed with sand by the waves and currents. Wind remodels the exposed parts, making dunes, and when plants get a chance to grow and anchor the sand, they freeze the dunes in place, until a big storm tears them up again. Barrier islands are long and narrow, extending along the coast. On the land side of these barrier islands is the Intracoastal Waterway, a low area which geologists refer to as a backbay. Rivers feeding freshwater into the back-bay are blocked from the sea by the barrier islands. This creates a marsh between islands and mainland. Eventually, the rivers cut tidal channels between islands, spilling their water into the ocean. The ocean fights back at high tide, pushing seawater into the breaks between the islands and overwhelming the backbay. Then, at low tide, the flow reverses again. Freshwater and seawater mix, and the saltiness of the water varies, from very salty near the tidal channels to hardly at all away from the channels. The flow of the rivers decreases greatly as it hits the backbays, and with that, it also loses the ability to carry sediment. The rivers carried mud, and the mud is dumped in the backbays, creating a rich, nurturing soil for the plants. Differences in salinity, water depth, and vegetation lead to multiple habitats, with very different animals in each. Driving over the bridge, I imagine an Eocene whale lurking in the mud, looking up at my car zipping by.

On the island, we grab bikes to go to the ocean. The island is so different from the backbay. Kiawah's soil is like beach sand, with lots of shells of clams and snails, animals that were alive when they were part of the seafloor but whose shells are now found in a different environment. The island is less than a mile wide, but it supports lots of wildlife. As we walk toward the beach, we see alligators. They consider this island theirs and are not afraid of humans. They bask in the sun in little ponds on the island, sometimes right behind the dunes. Those ponds collect rainwater and are not salty: these alligators live in freshwater.

The alligators remind me of the crocodile bones that we find in Kutch. They are often associated with seashells. I have always thought that these were bones of marine crocodiles, given that they are associated with the shells. I now realize that I should not rush to that inference. If a future paleontologist were to dig into Kiawah's soil, freshwater alligator bones would be both close to the ancient shore and associated with ocean clams, though the alligators did not live with the marine invertebrates.

We cross the dunes, dragging our bikes along, and find the bones of a gray fox, a mammal that hides in the daytime, in the dune vegetation. The fox probably hunts rodents and birds in the forested area right behind the dunes, part of the nocturnal fauna which this diurnal observer misses. On the beach there are lots of shells, all open halves. Storms have churned the shells of the dead molluscs, and thrown them on this beach, and it reminds me of Godhatad, one of our Indian fossil localities. On this beach, there are also sand dollars, flat relatives of sea urchins. In India, we always use sea urchins as indicators of an environment that is fully marine, since they do not tolerate freshwater. However, these sand dollars are only fifty yards from the alligator pond. This shore is a patchwork quilt, with very different patterns and colors in each of its squares, and I need to think of Kutch in the same way.

One of the reasons to come to this barrier island is to see the dolphins. Even though I have studied fossil whales for a long time and have dissected dead dolphins many times, I have never seen wild dolphins.

We ride our bikes on the beach to the southern tip of the island. Here the barrier island is separated by a tidal channel from the next barrier island. Water flows out of the channel fast. The tides drive this process. As the sea level rises around high tide, the water pours into the muddy flat areas behind the barrier island, flooding them with seawater through this channel. As the tide drops, all this water comes out again, as if a giant were lifting the mudflats and pouring the water back into the sea. Dolphins are smart animals. They know about the tide, and they know that with the pouring-out of the water many fish are poured out of the mudflats, too. The dolphins stay in the tidal channel, catching the fish that are forced through the channel.

Having an eight-year-old to please, I am worried that the dolphins might not be there.

"Do the dolphins always show up?" I ask the uniformed ranger. "It is a long trek for my son."

"The dolphins will be there," she says confidently, "right when the tide reverses."

We arrive a bit early for the show, and walk along the channel's edge. There are dozens of large snail shells: whelks. We collect them. They are large and beautiful—orange, yellow, tan, with darker gray blotches—and smell like they have been buried for a while. In fact, when alive, the whelks are buried in the tidal channel. They only become exposed when the animal dies and all the flesh is gone. It again strikes me that we didn't see whelks anywhere on the beach; they are only here, near the tidal channel. We pick up several, more than we can carry. For each bigger and more beautiful one that we see, we discard one of the earlier ones. We can't take a lot of whelks, because we have to bring them back on our bikes.

With my eyes on the ground, carrying the whelks in my arms, walking back and forth, I pay no attention to the channel until suddenly—*whoosh*. Loud and sudden, it startles me, and a bulging gray object the size of a basketball disappears. It was in the water thirty yards from me. A dolphin has arrived—its forehead was all I saw. I drop my whelks and sit in the sand. The ranger was right. They patrol the channel, coming up to breathe. The water is so muddy that I cannot see the body of even the closest one. But I can see the blowhole on the forehead. That is the only thing that comes out of the water—the eyes and ears are below the water-line—just that gray bulge. The Indian whales, too, might have hung out in the shallows, waiting to pick off fish that shot by in the muddy water.

We watch the dolphins for half an hour. The sun is setting. The marshes turn the orange of the whelks in my pile, and then shades of gray, as we bike home.

I think back to Kutch. At one site, there may be one hill with gypsum and sirenian ribs, and another hill consisting of broken oyster shells. All represent different fossilized environments, a stone's throw from each other. There, too, the orange sun sets over the Harudi Formation, as it did forty-two million years ago, when the Harudi was the coast.

A FOSSILIZED COAST

The Kutchi fossil localities stretch along a C-shaped band about seventy miles long, circling a central area of exposed land in the Eocene (figure 30). There was a diversity of habitats along this band. Just as in modern times, the Indian Ocean was to the south of the localities, and a large sea arm extended around the peninsula to the west and north. Nowadays, that arm is dry most of the year and is called the Rann of Kutch, but in monsoon time it fills with water and turns the Kutchi desert soggy.

In the Eocene, the southern fringe of the localities was closest to the ocean. The fossil locality Rato Nala is in this area, and large algal mats with molluscs dominated it, implying shallow and clear water. The algae precipitated calcium carbonate and thus constructed an algal reef that fossilized as the Chocolate Limestone. Molluscs that were living in the algae were entombed and smothered by them, and dead whales and sirenians sank to the bottom, where they were also encased by algae—the minerals released from the dead mammals feeding the algae. But that was just one environment at Rato Nala. There were also muddy shallows with lots of small water plants, now recognizable as plant fossils. Then there are gray muds with veins of yellow sulfur, apparently formed in an anoxic environment.

Fifteen miles east of Rato Nala, also on the south side of the Eocene land, is the locality of Vaghapadar. Here are lots of enormous marine snails and sirenians, but relatively few whales. Sirenians are excellent indicators for fossil environments: they are aquatic plant eaters, specializing in seagrass. In the Eocene, Vaghapadar was probably a seagrass meadow, too suffocating for the fast-swimming whales but just right for the slow-grazing sirenians.

Moving north along the ancient shoreline, into the sea arm, is the locality Godhatad that I thought of while visiting Kiawah Island. Being farther from the open ocean, it was protected from the waves and

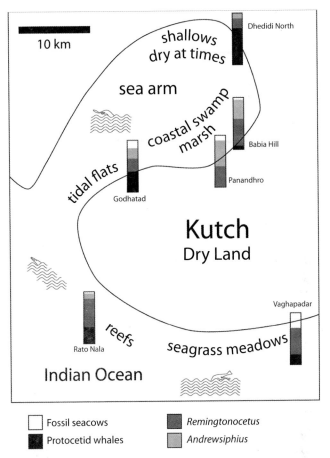

FIGURE 30. Map of western Kutch (India) in the Eocene, with fossil localities and the environments occurring there. Different kinds of marine mammals prefer different habitats, and this is reflected in their abundance at different fossil localities. The white-gray-black bars show the proportions of seacows and the different whales.

weather out on the ocean, but it remained connected with that ocean. Sediments and fossil plants indicate that Godhatad was composed of tidal flats and lagoons in the Eocene.[1] It is also a patchwork of environments, and the most impressive of these is a long hill composed almost entirely of broken oyster shells. In the Eocene, it was a storm deposit: the storm killed the oysters and smashed their shells, dumping them in large piles. Even though there are conspicuous molluscs at both places, the difference from the molluscs of the Chocolate Limestone is striking.

At Godhatad, there are no oysters with their shells closed. Those in the Chocolate Limestone are mostly closed, indicating that they were buried alive. At Godhatad, fossil whales are found, too; they were either killed in the same storm, or maybe the storm moved dead whale parts around, burying them with the oysters, to the delight of paleontologists more than forty million years later.

Northeast of Godhatad are large deposits of lignite, a poor-quality coal. These deposits were formed in a salty swamp or marsh[2] and can be seen in the Panandhro Lignite Mine and Babia Hill localities. Plants died, and did not rot; instead, the plant debris was covered by more plants and buried. Anaerobic conditions were common, as evidenced by the formation of pyrite, and a sulfurous smell when you hit the rocks with a hammer. Panandhro was sheltered from the churning of the ocean by the land to the south. It represents a forested swamp with stagnant water in which many whales lived.

Farther to the north, and farthest from the ocean, is the locality Dhedidi North. Gypsum abounds here and was formed by the drying up of a saltwater basin. Many of the fossils are covered in gypsum, suggesting that whales here died as their lagoon or bay dried up. That process continues into the present. When the Rann of Kutch dries up in summer, it produces enormous salt deposits visible from space.

Applying the lesson from Kiawah to Kutch, I can now imagine being here forty-two million years ago. Along a few miles of coastline, there are lots of different environments, with different plants and invertebrates, and different mammals, too. Whales that are common at one place are rare at another. The sedimentology can teach me about the whale habitats, and that in turn can teach me about what the whales needed to live.

CHAPTER 8

The Otter Whale

THE WHALE WITH NO HANDS

Kutch, India, January 12, 2000. The desert of Kutch is mostly uninhabited, except for a few herders, who roam the plain with their flocks. However, there is that one place that is teeming with humans who are not pastoralists: the lignite mine at Panandhro. It is a giant open-pit mine, one of the largest in India. Enormous machines make you feel the way an ant must feel standing next to a blender—awed by the size, but puzzled by the function. Hundreds of people work there, and the mining company built a town for them and their families, as the local villages could not sustain so many people. This company town—the Colony, they call it—has straight streets and identical houses, a shopping center, a school, a playground, and a desalination plant for seawater. Busses take workers to the mine, identical white SUVs move the engineers around, and they bark commands at the colorful trucks that are loaded with lignite and that clog the road all the way to the nearest real town, about three hours away. By the grace of the mining people, we're allowed to stay at the guesthouse. This solves many of our logistical problems: in a desert, where do you eat, where do you get water, where do you buy supplies?

In the morning we drive to our field area. Little Indian antelope, the size of goats but more nimble, cross the road, and we lose centuries as we pass the occasional villages with no cars or paved roads, water buffaloes lingering in ponds and used by boys as diving platforms, while their sisters carry jugs of water to their houses.

As we approach the village of Godhatad, I spot a small dog-like creature, a jackal, crossing the road. "It is mad, sir," states the driver, matter-of-factly.

I assume he infers this from the animal's being out at this time of day. But it bothers me. We, too, will be out here all day, far from the car, looking for fossils. Does the clinic provide rabies shots in case someone gets bitten? Where is there a clinic, anyway? Does the Colony have one?

From up on a ledge of the yellow Fulra Formation, I look down on Godhatad. Its twenty or so houses share walls, or verandas, or lean into each other, as if they are hugging, huddled together to keep the heat out. They are painted a light purplish blue—periwinkle, my assistant calls it. The periwinkle stands out against the drab rocks around it. Godhatad is a Muslim village; half a mile away is a similar Hindu village. Not that I can tell. I don't see a mosque, and I don't see any people from where I am. Also, the Muslims here mostly do not dress differently from the Hindus. The women do not cover their faces. They wear the same bright reds and oranges as their Hindu counterparts, a striking contrast with the drab and periwinkle surroundings. And nearly everybody, men, women, Muslim and Hindu, cover their head, to keep sun and dust out.

We are looking for a track across a valley that the car can work its way through. We drive around the village, but there are no clear tracks that descend the steep escarpment to the place where we need to go. Little kids have come out to watch. Except for the government truck that brings drinking water and the beat-up bus that takes people to town once or twice a day, cars rarely reach here, and white people are as common as rain in July. The kids follow the truck, which moves "dead slow," as they call it in India, searching its way over the bumpy cattle track. The driver decides to drive into the village and ask. The passages between houses are narrow, made for bullock carts, and our truck is barely able to make the turns. Hearing a vehicle, people come out, heads around walls, eyes over fences. The driver calls out to someone, and a man with two heavy golden studs as earrings approaches. They talk. I get out a bag of candy, always with me for this particular purpose. I hold a piece of candy up to the kids and wave. Three boys approach, ages six to nine, I guess. I give them each a candy. Broad smiles. One takes off to show his family, the others stay. Some girls stand further away. They do not dare to come close. I motion one of the boys to the candies still in my hand and point at the girls, but he doesn't get that I want him to take the candies to the girls. I don't dare to wave the girls over here. This is a Muslim village, and one of them is at least nine or ten. In Pakistan, it would be inappropriate for a stranger to talk to a girl that age, and I am not sure how people in Kutch feel about that. I feel helpless. My companions are now all deeply involved with the directions, so I cannot ask them to help. An older woman stands near. I lift my candy hand in the direction of the girls and ask her, using my very limited Hindi, "Teek-haeh?" (OK?) Her weathered and wrinkled face looks friendly but uncomprehending. This is frustrating.

Everybody else packs back into the car. I need to leave, too. Too bad, no candy for the girls today. I give a handful to one of the boys, point at the girls, and say, "Please share." He smiles again. Has no idea what I want. Maybe his mother will make him share with his sisters and cousins, or maybe he'll eat the lot himself and get a tummy ache.

Back in the car, we cross into a low-lying bushy area and park the car. We walk down through a nala with a muddy bottom, very unusual in Kutch. The dense bushes are taller than I am, and there are rustling and grunts from that area.

"Wild pigs," the driver informs me casually, chewing on a blade of grass.

I look up. Those are scary, very aggressive when they have babies. I don't know if they have babies this time of the year, and I do not want to find out. The grunts in the underbrush are much too close for comfort. I look at my hammer. It seems an insufficient weapon against a mother pig that feels threatened. I rush and quickly walk across the wooded area to where the desert takes over again. I like desert—you see the rocks that have the fossils, and you see the mad dogs and hogs before they are upon you.

I look back at the hog-infested nala, and across to Godhatad. The peace of the desert and people of India surround me again. The azure cloudless sky, yellow limestone cliff background, the periwinkle village with its red tile roofs, its little lake where the water buffaloes are cooling off, and the green nala, set against the drab hills. Fossils or not, I love it.

At Godhatad, there is a lot of Harudi Formation exposed, including the large storm deposit of broken oyster shells. We walk the oyster bed, eyes on the ground, looking for fossils. I put my backpack near that of Ellen, the fossil preparator, who is collecting higher on the same slope. The backpack is heavy with a gallon of drinking water—Kutch is hot—and then there is lunch, a chisel, glues, dental tools, brushes, knives, and plaster for encasing fossils for transport. Sunil has put his bag in the shade of a rock. I never do that, after having once spent an hour trying to find my backpack that was too well hidden in the shadow of a rock. A hot backpack is better than no backpack.

Suddenly, I find a fossil, a piece of a vertebra. Not very nice, but a good start. I follow the gulley that it was in and run into another vertebra. Ellen comes and helps. She finds another one, and another one. Then Sunil appears. "Hans, I have found something."

I am slightly annoyed that he expects me to stop what I am doing and come over.

"Is it any good? I am having some luck down here."

"There are many bones."

"Are they identifiable?"

He smiles. "You will see."

I retrieve my backpack, and we go to a flat area barely a hundred feet away. Pieces of vertebrae are strewn across the area here. There is also a bunch of pieces of bone, mostly smaller than a dime. We call them chips. But no nice, big and complete things.

"Lots of bone, many vertebrae. Did you find anything else here?" I ask.

He shows me the fossils in his pocket: mostly pieces of vertebrae, some small parts of bones too fragmentary to recognize, except for a piece of a femur. One vertebra is from the tail, and it is more than twice as long as the other vertebrae he has. Interesting. I try to imagine the animal. Very long tail, not like *Ambulocetus,* more like a basilosaurid. Could it be a new whale?

"Sunil—nice vertebrae, lots of them too, but where are all the long bones, and the skull?"

At this point, there is not even enough to tell that this was a whale. We need to collect harder here, maybe excavate this place.

Ellen and I walk around in widening circles, figuring out where the greatest concentration of bone is. Then we put large rocks around the area with the bones to mark it. We divide the area inside it into segments and crawl over each one, picking up every piece of bone. We find two vertebrae still partly buried, and a larger fossil too. We do not touch the buried ones, planning to excavate them later, once the loose surface finds are in the bag. The little piles grow bigger, but there are just vertebrae and unrecognizable fragments. I will not be able to tell what this animal is. I have visions of my previous experience in Pakistan. With *Ambulocetus* it took four frustrating days before the skull was found and the beast became identifiable. Except for that half femur, we don't even have limb parts after an hour. Knee pads would also be nice—the entire surface of the ground is covered by little rocks the size of gravel. Geologists call this "desert pavement." It is formed when wind blows away the fine material, leaving gravel and rocks behind. Eventually, all the fine material is blown away, and those big pieces cover up the entire surface, keeping the wind from eroding the ground further. Desert pavement is very hard on pants and knees, and eventually on the motivation to collect.

These dozen or so vertebrae with chips definitely have potential. The tail vertebrae all seem to be large. Ellen and I now turn to the area with the concentration of still-buried bones. She digs and exposes more vertebrae, and I work on a single large piece of bone that is also buried. Ellen loosens dirt and gravel with a dental tool and brushes it away with toothbrush and paintbrush, exposing fossils buried deeper, leaving tiny fossils on tiny pedestals. More vertebrae emerge. I would love to have the entire vertebral column for this beast, but I would love even more to know what it is. Another hour passes, and the little excavation is a few inches deep. Sunil has taken off looking for other fossils. The desert is quiet and hot.

In silence, Ellen and I continue. The big fossil is impressive—it is not a fossil exactly, more like a big impression of bone in the rock, with a few little bone pieces the size of a fingernail still stuck to it. The impression has the shape of a Y, with a very long stem. I hope that it is connected to something nice underneath. I cannot figure out what it is, or rather what it was before it all eroded away. I lie back, stretching. My back hurts from hunching over. Ellen breaks the silence. "What is it?" She has noticed that this bone is irritating me.

"Nothing very nice. This bone bothers me. I don't know what it is. It is too long to be a limb bone or a vertebra."

"Part of the skull?"

"I don't see it. It can't be the braincase. It can't be a snout. It can't be a jaw, either—those do not split that far back. I had hoped to see teeth or alveoli."

In my head, I run through the entire anatomy of a mammal over and over again, trying to fit this silly big-Y thing somewhere. I cannot figure it out. Instead, I take its picture and try to put it out of my head. "Photo of skull fragment, very bad, not collected," I write in my notebook, and leave it.

Sunil comes back with a large rib, a sirenian, but a big one. Its rib is the size and shape of a banana. I walk over to the spot with him, leaving Ellen behind. She continues to excavate. The rib is a bust. The rest of the sirenian is not there, and I return to Ellen an hour later. She has produced a large pile of chips, most of which are too small to identify. The hole she dug is now two feet wide, and none of the chips she has found are part of a recognizable bone. I am disappointed, and write in my notebook, "Three hours later there is no improvement."

I am getting tired of this. "Let's cover some more volume using the pickax. We can excavate deeper, and if it is just chips we're getting, it is not like we'll be destroying anything."

We get the pick, and I swing it to loosen the rocks. After two hits, Ellen sorts through the loose dirt with her hands to look for fossils. She finds a bigger piece. It is a piece of the humerus. That is important. It turns the tide, and my mood.

"Wonderful, a limb bone." Few limb bones are known for these Indian whales, and none are associated with vertebrae. Of course, if it indeed is a whale, we need to find teeth or skull. I hit the soil again, near where the bone came from. The pick cracks. Hands sift through loose sand and rocks.

"Wow, a distal tibia, another long bone. Maybe there is a complete skeleton here."

We now become absorbed by the digging, and find more pieces of limb bones, smaller ones, but still. Hours pass. The sun throws long shadows. Sunil returns to the site, and I summarize.

"So, what we know is that this animal had short, sturdy, squat limbs, especially the parts of the limbs close to the body. Powerful swimmers, diggers, and climbers have that, making their limbs into levers to move in a dense medium. It also had a big, strong, and long tail." I hold a tail vertebra next to the humerus. The bone is barely twice the length of the vertebra.

"Amazing—this guy was mostly tail, with short stubby legs. This is going to tell us a lot about locomotion in this beast. Powerful tail, for sure."

"So, what is it?" Sunil's comment brings my musings about swimming back to ground level.

"Right. We have to find out what this beast is."

We only have the parts of the limbs close to the body, nothing below wrist and ankle. I decide that we should excavate it more, and sift all the sediments through a screen, to find the smaller pieces, like the hand and foot bones.

It is time to go. We wrap our bones in toilet paper, like we are making little white gift packages. Really, this is better than Christmas.

Driving to the Colony, we make plans. We will sieve the sediment so that dirt and tiny rocks will fall through but everything larger than a pea will stay on the sieve and can be sorted through quickly. We stop at the small bazaar next to the Colony and go to the hardware store. The store is the size of a large bathroom and packed to the roof with shovels, and rolls of wire, but also with rolling pins and griddles. Sunil translates my description of what I need. There is a lot of back and forth, but the man seems competent and eager, and I leave the store with great confidence in the Indian hardware industry.

In the evening, we unpack the fossils. Ellen joins the ant colony that calls our bathroom home, and washes the fossils in the sink. Now, with clean fossils, several chips snap right on to the humerus and femur, making two complete limb bones. Very satisfying!

Ellen also exposes some dime-sized pieces that are dark gray, not tan like the other bones. They are enamel—these are parts of teeth—whale teeth! This is a whale with legs and a strong tail, certainly very different from *Ambulocetus* and *Basilosaurus*. A new species and clearly an awesome find.

The next evening we pick up our screen at the hardware store. The owner meets us outside—the screen is too big to fit in his cluttered store, and much bigger than I had imagined. It now occurs to me that the dimensions I specified mean that it won't even fit in the car. Also, it is not a screen. Instead, the maker has nailed a piece of sheet metal to the frame, and punched hundreds of small holes in it—an ingenious solution to the lack of screen at the Colony. The entire thing costs 45 rupees, about a dollar.

The driver solves my problem by tying the screen to the roof of the car. Back at the Godhatad site, I dump three shovel loads of sediment into our sieve. Ellen and I each grab a side and shake it rhythmically. It is heavy and dusty work. The shaking throws us off balance; dirt gets in our boots, and the wind blows it into our eyes. We find fewer pieces than I had hoped. I forgot my hat, and my ears are burning. Ellen loans me hers for a while. My voice is still cracking from a cold I got on the plane, so I eat a lot of cough drops. Sunil calls me "delicate."

The evening ritual of washing and fitting repeats itself. More vertebrae and pieces of long bones, but nothing really nice. One of the vertebrae is nearly complete, but its end has a triangular piece missing. Another vertebra has an odd triangular lump sticking out. I fit the two together and to my shock I realize that they go together. These vertebrae were fused in life. Ellen sees me do it. "A sacrum?" she asks.

"Yeah! But where is the rest of it?"

We both frantically search for more in the bag of pieces.

Geological age-wise, our new whale falls between *Ambulocetus* and the basilosaurids. The sacrum of those two is very different: four firmly fused vertebrae in the former, no fused sacral vertebrae at all in the latter. That makes the difference between being able to support one's weight on land and not, so this is very important in understanding our new whale. Ellen and I keep on going through our fossils, finding more pieces. I glue them onto the growing sacrum with white Elmer's glue. It takes a while to dry, and I am impatient, looking and trying to fit more fragments before already-glued joints have dried. Some come apart from my handling the fossil. Ellen notices my impatience and takes the sacrum from me without saying anything, firmly but gently. I know better than to dissent. She is the fossil preparator. Things progress more slowly, but now we need to glue each joint only once. Eventually, she fits a nearly complete sacrum together: four fused vertebrae, with a large joint for the pelvis. This guy could certainly stand on land.

Still high on the success of our reconstructive surgery on the sacrum, I keep sorting through our bags of fossil chips. My eye is drawn toward a bone the size of a toffee, and also that color. It is broken on three of its four sides, indicating that it was part of something bigger, and the breakage also explains why I ignored it before. However, the fourth side shows two holes; they're for teeth! A shiver runs over my back. "Sunil, I've got a jaw."

He looks up. The cavities that hold the left and right teeth, the alveoli, are very close together, indicating that the jaw is very narrow. We both know what this means: this is one of the narrow-snouted whales for which we have fragmentary skulls. It allows us to identify the specimen as a remingtonocetid.

Eventually, this new whale will receive the name *Kutchicetus minimus*, the "smallest whale from Kutch" (figure 31).[1]

Over time, too, it becomes clear what the big Y was: the impression of the underside of the lower jaw. The long stem is the part where the left and right jaw touch; the short arms are the left and right parts as they diverge. The flute look-alike from Babia Hill, mentioned in chapter 6, would make just such an impression, and, indeed, they represent the same species.

As *Kutchicetus* becomes well known, a company that makes museum exhibits, Research Casting International, puts it all together and casts it in some fancy plastic, for use in museum displays. Peter May, its

FIGURE 31. Life reconstruction of the Eocene whale *Kutchicetus minimus,* which lived in India around forty-two million years ago. *Kutchicetus* and other remingtonocetids were probably fish eaters, and were able to walk around on land.

director, and I get all the bones together, and his team makes mirror images of the bones for which we only have one side. For the feet, there is nothing to mirror image because we found no fossils of them. I do not want the feet reconstructed, because I do not know what they looked like. We decide that he will use wire to indicate where the toes were, thus leaving it very clear that we do not have those parts.

Then I ask Carl Buell, a scientific illustrator, to make a drawing of the animal. I know Carl. He is picky and precise. He asks for details—this view, that bone, things that I have never thought about.

"How are the lips, are they floppy like a dog, or tight like a whale?" Carl asks in his crackling voice that betrays that he is not a young man. Carl knows his stuff—anatomy, function—he is also passionate. I say that I do not know the answer to his question.

"It has this long, narrow snout, that's crazy, I'd love to wrap my pencils around that—a whale looking like a gharial." He is right. This whale does look like the narrow-snouted crocodilians that inhabit India and Pakistan.

"So what do its feet look like?"

"We don't have the feet. We didn't find them."

It is quiet for a second. He's clearly disappointed.

"You don't have any bones from the feet—nothing, no phalanges, no carpals, nothing?"

"I have half a bone, probably a phalanx, it won't help you."

Quiet again. I break the silence. "What are you going to do with it in the reconstruction?" I ask, somewhat worried that he may not think the project is feasible.

"Oh, I'll figure out something." This does not sound good. I want to know what is he planning.

"I don't want you to make them up, OK?"

"You'll see, you'll like it."

His voice changes to the tone of a doctor who comforts a worried patient, but does not want to explain. I trust Carl, so I don't push him. He wants to know what to use as a scale, so people will know how small it was. He proposes a shorebird. People have some sense of what size those are, and Carl knows that they were around in the Eocene.

A few days later, Carl sends me some sketches, the head from dorsal and from the side—he is clearly struggling with it, it is so different from other mammals—but no reconstruction of the body. I still do not know how he is going to solve the hand-and-foot problem, and I am dying to know. Then, eventually, I get a sketch of the entire animal. I rush to open the file. There are three individuals, showing three different views of the head, and he's put them at the shoreline, feet in the water, invisible, the rest of the animal above the water, visible. Brilliant, such a simple solution, except that I would not have thought of it. The bird in the background looks good, too.

REMINGTONOCETID WHALES

Carl's reconstruction put some flesh on the bones of the remingtonocetid whales, initially discovered by Ashok Sahni and his student V. P. Mishra. After that initial description,[2] Sahni sent another student, Kishor Kumar, to Kutch to collect more whales. Bad weather made fieldwork impossible for much of Kishor's stay, but he did find the most complete skull of *Remingtonocetus* known at that time.[3] He also collected new material for another whale that Mishra had discovered: *Andrewsiphius,* named after C. W. Andrews, a British paleontologist who worked in Egypt and described many basilosaurids. Realizing that *Andrewsiphius* and *Remingtonocetus* were part of a unique Indo-Pakistani radiation of whales, Kumar and Sahni combined *Remingtonocetus* and *Andrewsiphius* into a new family: Remingtonocetidae. Since then, no remingtonocetid has ever been found outside of the Indian continent, but three additional genera have been described: *Dalanistes,* based on specimens from central Pakistan[4] and Kutch,[5] *Kutchicetus* from Kutch, and *Attockicetus,* which came from the same rocks as *Ambulocetus:* the Kuldana Formation of Northern Pakistan.[6]

Remingtonocetids differ from other Eocene cetaceans in having long snouts, tiny eyes, and big ears. Based on what is known for *Kutchicetus,* its body looked like that of an otter: short legs, and a long, powerful tail. In contrast, the long, narrow snout makes the head look more like a gharial (figure 33). *Kutchicetus* was the smallest remingtonocetid, the size of a sea otter; *Dalanistes* was the largest, weighing maybe as much as a male sea lion. Indian remingtonocetids are known from rocks forty-two million years old.[7] *Attockicetus* is older, the same age as *Ambulocetus,* approximately forty-eight million years, and the remingtonocetids from central Pakistan are between thirty-eight and forty-eight million years old.[8]

Feeding and Diet. It is easy to imagine that the long snout helped remingtonocetids catch fish. If *Ambulocetus* lived like a crocodile, capturing large, struggling prey, remingtonocetids were more delicate, lashing out quickly with their sharp teeth when a fish came close.[9] The front teeth of *Kutchicetus* are long and slender, good for piercing and checking slippery prey in a dash, but not for holding powerful struggling prey. The molars are small, but tooth wear shows that *Remingtonocetus* chewed its food, unlike modern whales; its teeth worked like those of basilosaurids and ambulocetids. These molars cut like scissors, with sharp shearing edges

FIGURE 32. Life reconstruction of the remingtonocetid whale *Kutchicetus*. Fossils of hands and feet were not discovered for this whale, which means that the artist reconstructing the animal needs to be creative.

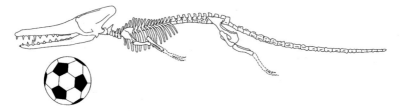

FIGURE 33. The skeleton of the Eocene whale *Kutchicetus minimus*. Soccer ball is 22 cm (8.5 inches) in diameter.

(figure 34). No part of these teeth is involved in crushing food, unlike *Ambulocetus*. Analysis of the stable isotopes of the teeth is consistent with a fish diet, and further study may refine this.

The flute look-alike mentioned in chapter 6 is a *Kutchicetus* jaw from Babia Hill. All its teeth fell out after death but before burial,[10] but counting sockets for the teeth (the alveoli) reveals what the dental formula is. As in most early whales, there are three incisors, one canine, four premolars, and three molars in both upper and lower jaw: 3.1.4.3/3.1.4.3. Some jaws for *Remingtonocetus* and *Dalanistes* do still have teeth, and surprisingly, the lower molars look like those of basilosaurid whales (figure 34), with multiple cusps of decreasing size lined up from front to back.[11] However, these teeth are unlike basilosaurids' in that they are slender and delicate, not built to mince tough or hard food. In *Andrewsiphius*, there are three low and flat cusps on a lower molar, lined up in a row, with the middle cusp barely higher than the others. It is not clear whether this unusual shape somehow related to a specialized function.

Remingtonocetid molars are specialized, unlike those of ambulocetids, which retain the shape of archaic land mammals with a high front (the trigonid) and a low back (the talonid). The premolars in remingtonocetids have simple, triangular cusps. In *Andrewsiphius* and *Remingtonocetus*, most premolars have two roots, but in *Kutchicetus* there is only one root per tooth. In modern whales, there is never more than a single root per tooth.

The most unusual feature of the jaws of remingtonocetids is the long area of contact between left and right lower jaw. This area is called the mandibular symphysis (figure 25). In ambulocetids and basilosaurids, the left and right lower jaws are connected by ligaments; there is no bony fusion across the mandibular symphysis. This is also the case in most *Remingtonocetus*, although there is a bony connection in old individuals. In *Andrewsipius* and *Kutchicetus,* the left and right jaws are joined by

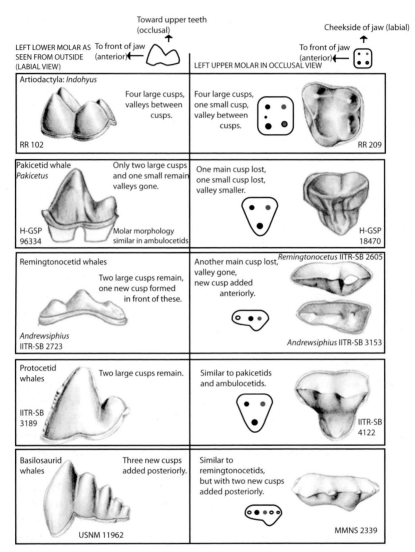

FIGURE 34. Left lower (left column) and upper (right column) molars of Eocene whales and the artiodactyl *Indohyus*, showing the vast differences in the topography of teeth. Outline diagrams for upper molars show how the position of cusps changes in evolutionary time. *Indohyus* is discussed in chapter 14.

bone: this is called a fused symphysis. An unfused symphysis allows for some independent movement of the jaws during chewing, and most small mammals have unfused jaws, for instance dogs, cats, rabbits, and rats. Larger animals that eat plants, such as horses, cows, elephants, rhinos, and hippos, tend to have fused symphyses. Large, long-snouted animals, such as crocodiles, also tend to have fused symphyses. The long symphysis, fused or unfused, gives the jaw strength when it is slapped shut, possibly preventing teeth from interlocking incorrectly (misocclusion).

Breathing and Swallowing. The nose opening of remingtonocetids is near the tip of the snout, which is where it is in land mammals. With the long snout, it may have allowed the whales to lie in wait in deeper water, with the tip of their nose above the water, and thus avoid the need to come to the surface to breathe while hunting. However, a nose opening that far forward might be helpful for other reasons. For remingtonocetids, freshwater conservation, in the face of the salty ocean they lived in, could be important. Modern seals use their nasal cavity to retrieve water from exhalation: water vapor coming from the lungs condenses in the nasal cavity and is taken up by the tissues inside the nose.[12] It is possible that the long nasal cavity in remingtonocetids served this function.

The remainder of the skull has its peculiarities too. The hard palate of *Remingtonocetus* extends nearly to the area of the ears, as in *Ambulocetus,* although it does not reach as far down (ventral) as in *Ambulocetus.* There is a prominent midline crest on the hard palate, probably for the attachment of the chewing muscles that attach to its lateral side: the left and right medial pterygoid muscles. The medial pterygoid is a powerful mouth-closing muscle, useful in a fish eater that clamps its teeth down fast on prey. In closing the mouth, the medial pterygoid works with two other muscles: masseter and temporalis. Temporalis attaches to the side of the skull, as well as to a crest on top of the skull (the sagittal crest, figure 29), and those bony attachments suggest that it was a very large muscle in *Remingtonocetus* and *Andrewsiphius.* Masseter attaches to the outside of the jaw and to the jugal arch, formed partly by the cheekbone. The jugal arch is tiny in *Remingtonocetus,* suggesting that masseter was small. This is strange because, in most mammals, medial pterygoid and masseter are similar in size as they work together closing the jaw. As in other mammals, the throat anatomy of *Remingtonocetus* combined the functions of chewing, swallowing, and breathing, but the interplay of these is poorly understood in this fossil whale—but clearly different from *Ambulocetus.*

Brain and Smell. The skull of the *Remingtonocetus* specimen in figure 29 is an unusually well-preserved fossil, and it was possible to CT scan the specimen and study the outside as well as the inside cavities of the skull, such as the nasal cavity and the braincase. A sophisticated scanner took about a thousand scans, covering every half-millimeter of the specimen (figure 35). A special computer program reads those slices and puts them together, so that pieces can be left off or added again, making virtual removal of parts possible. In figure 35, the skull is left off, and just the cavities inside it are shown—a virtual endocast (as described in chapter 2). The CT scans show that the brain was small, and on its sides are large areas that were probably bunches of veins, as in the modern whales discussed in chapter 2. As in all mammals, the outside of the brain consists of two large parts: the cerebrum in the front, and the smaller cerebellum in the back; both can be seen in the virtual endocast of figure 35. Very unusual is the canal that emerges from underneath the cerebrum in the front and reaches toward the nasal cavity. In life, the nerves going to the nose (cranial nerve I) would have run in this canal, but usually this nerve is not as long as it is in *Remingtonocetus*. The contact of the canal with the nasal cavity does indicate that *Remingtonocetus* had a sense of smell. However, it is not clear why it is as long as it is. At present, it seems most likely that the external anatomy of the skull affected skull architecture and thereby internal anatomy: the position of the masticatory muscles on the outside of the skull in this area required that the internal structures be elongated.

Vision and Hearing. The position of the eyes is unusual in remingtonocetids. They face laterally underneath a dome-shaped forehead in *Remingtonocetus, Dalanistes,* and *Attockicetus* and are not perched on top of the head. The eyes in *Andrewsiphius* and *Kutchicetus* also face laterally, but are closer together on top of the head, more like *Ambulocetus*.[13] The small sockets for the eyes suggest that the remingtonocetids had poor vision. These animals lived in muddy water and a swampy environment, so probably there was not much to see. In contrast, the ears are enormous: the two large tympanic bones that surround the middle ear cavity project prominently from the base of the skull. Of course, being whales, these bones do have an involucrum. Another indication that hearing is important is that the mandibular foramen is nearly as large as the full depth of the jaw (figure 25), similar to modern toothed whales, but bigger than in *Ambulocetus*. Some of the specimens from Kutch include ear ossicles, and they look very much like those of modern

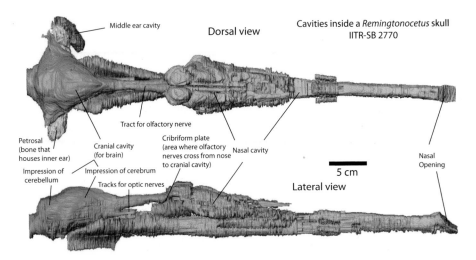

Middle ear cavity

Dorsal view

Cavities inside a *Remingtonocetus* skull
IITR-SB 2770

Tract for olfactory nerve

Petrosal
(bone that
houses inner ear)

Cranial cavity
(for brain)

Cribriform plate
(area where olfactory
nerves cross from nose
to cranial cavity)

Nasal cavity

Nasal
Opening

Impression of
cerebellum

Impression of cerebrum

Tracks for optic nerves

5 cm

Lateral view

FIGURE 35. Internal anatomy of the skull of *Remingtonocetus* (IITR-SB 2770) as reconstructed from CT scans. The cranial cavity (where the brain is located) is green, the nasal cavity with its sinuses is blue, and the middle ear cavity is red. The bone that houses the organs of hearing and balance (yellow) is called the petrosal (see chapter 11). After S. Bajpai, S., J. G. M. Thewissen, and R.W. Conley, "Cranial Anatomy of Middle Eocene *Remingtonocetus* (Cetacea, Mammalia)," *Journal of Paleontology* 85 (2011): 703–18. Used with permission of the Paleontological Society.

whales, and unlike *Pakicetus*. Clearly, hearing is an organ system that is undergoing fast evolutionary change, and this is discussed in chapter 11.

Walking and Swimming. Among modern mammals, the skeleton of remingtonocetids is most similar to that of otters, suggesting that these whales were probably agile hunters. In Frank Fish's concept of the evolution of otter locomotion (figure 20), *Kutchicetus* would have matched the giant freshwater otter *Pteronura,* swimming with its powerful tail, possibly aided by paddling with the hind limbs (since the feet are not known in *Kutchicetus,* we cannot be sure). The vertebral skeleton for *Remingtonocetus* shows that it had a relatively stiff back, possibly more stiff than *Kutchicetus,* and it is possible that the species was a pelvic paddler;[14] but since neither tail nor feet are known for *Remingtonocetus,* locomotor inferences are speculative. In general, remingtonocetids were probably adept swimmers, but land locomotion must have been clumsy.[15]

Life History and Habitat. Remingtonocetids in Kutch are known from nearly all fossil sites there: the algal reef of Rato Nala, the seagrass meadow of Vaghapadar, the muddy storm-swept beach at Godhatad, Panandhro's swamp, and the dried-up sea arm near Dhedidi (figure 30). *Remingtonocetus* is common at all localities, and thus apparently not picky about its particular environment. *Andrewsiphius* and *Kutchicetus,* on the other hand, are rare at the localities open to the ocean (Rato Nala and Vaghapadar) but common at the localities that have muddier water and restricted flow, like Panandhro and Dhedidi. They appear to have been muddy-water specialists.

BUILDING A BEAST OUT OF BONES

It pleased me that Carl Buell did not make up the feet of *Kutchicetus.* We do not know enough about those feet to guess at their shape. On the other hand, I do not have a problem with Carl giving the animal brown fur, even though I have no idea what color it really was and can only make an educated guess that the animal had fur at all. Reconstructions are useful because they give an interested audience an intuitive feel for an animal—what it looked like and how it lived. Laypeople are unlikely to notice such details as how many toes there are, so artistic license in those areas does not violate the trust between scientist, artist, and reader. Of course, there is always some level of conjecture in reconstructions. If zebras were extinct and horses were not, it would be straightforward to draw a zebra's body shape accurately from its bones, but it is unlikely that any artist would get the color pattern right.

There is much disagreement among paleontologists about how far you can go in reconstructing an animal known from some bones only. In whale artistry, *Pakicetus* became a household concept in paleontology labs when it was described in the early 1980s. At that point, only a lower jaw, a braincase, and a few isolated teeth were known, but the cover of the prestigious weekly *Science* showed the animal jumping out of the water, with head, body, feet, and tail drawn in detail. Although the paper that described the animal was explicit about what was known, those nuances were lost in the many spin-offs based on the *Science* cover, including in reams of popular books and illustrations at natural-history museums. Such excesses of artistic license have not gone unnoticed in the creationist community, and have been exposed as examples of evolutionists making things up based on "a few scraps of bone."[16]

CHAPTER 9

The Ocean Is a Desert

FORENSIC PALEONTOLOGY

In the Del Rio, a Bar in Ann Arbor, Michigan, fall of 1992. My friend Lois Roe and I are graduate students talking shop at a bar. She went to Pakistan to collect fossil fish from the time that the Himalayas were rising, around fifteen to five million years ago, but they did not find many fossils. Now she is exploring questions that are less dependent on having many fossils, to get the most out of the samples she does have. She now works with a professor who knows very little about fish but a lot about the chemistry of rock and bones—an isotope geochemist.

Isotope geochemistry is a hot field of research, and can tackle some remarkable problems. It studies the subtle differences between different forms (isotopes) of the same chemical element. Oxygen, for instance, occurs most commonly in one form, ^{16}O, where the 16 indicates the weight of the oxygen atom. In nature, there is also a heavier isotope, ^{18}O, which carries two extra neutrons in its nucleus. Both isotopes react identically with other elements. For instance, $H_2{}^{16}O$ is a water molecule with a ^{16}O as its oxygen, and this is what most water molecules in nature are. However, there is also a little bit of $H_2{}^{18}O$ in the world. These are *stable* isotopes, meaning that they do not decay: once around they do not change, and they do not produce radioactivity. This is different from *radiogenic* isotopes, like those of uranium.

"In nature, ^{18}O makes up about 0.2 percent of the oxygen. There is no chemical difference, but the isotopes differ in their physical properties," Lois explains, while I sip my beer.

"Like what?"

"They fractionate according to their physical properties."

"What is fractionation?" I know nothing about isotope geochemistry, but I am not self-conscious about that with Lois.

"Physical processes will preferentially work with one of the isotopes. For instance, evaporation favors the lighter isotope—this can be used to track water through a system. I want to use this for my thesis work."

"Oh. Because the water molecules with ^{18}O are heavier, they evaporate less easily than those with ^{16}O, hence water vapor contains less $H_2^{18}O$ than the water in the ocean." It is now dawning on me what she means. If you can measure the ratio between ^{18}O and ^{16}O in water, you can determine whether the water you have came from water vapor or from the ocean. Since all freshwater eventually comes from precipitation, the difference holds for all freshwater too.

I say, "Those differences in the ratios must be tiny, and the weight difference between the isotopes is tiny too. Can you really measure that?"

"Sure, you use a mass spectrometer."

I know about the large machine in one of the labs across the street from the paleontology building. It vaguely reminds me of the top half of an enormous suit of armor, with two large metallic arms holding strange weapons stretching out of a larger irregular torso and head. The machine shoots out molecules from the hands of the knight, through its arms and into the chest, where they are deflected to different areas depending on their weight, and are counted when they crash somewhere inside the armor. The machine counts all those crashed oxygen atoms, and then determines the ratio of the isotopes.

"Cool, but what is that going to tell you about your fish?"

"Atmospheric water is fractionated as it moves up the Himalayas. The heavier isotope gets more and more rare because it rains out. Therefore, by measuring the isotope values, you can determine what the altitude was where the water sample was collected. Of course, you need to know the local geology, and the—"

"But you don't have water samples, you just have fossil fish bones. Where do you get the water?"

"The fish have drunk the water that they swam in, and used the oxygen in the water to build their bones. Bones are made out of apatite, which contains oxygen. Because the isotopes are chemically not different, you can measure the isotopes in the bones and determine the isotopes of the water they swam in."

"Wow. So the isotopes track the drinking water, and you can see what an animal drank, twenty million years after it died, and thus you can determine where in a river a fish lived: in the low plains, or the high mountain streams."

"The differences between different kinds of freshwater are relatively small, and they also depend on other things, such as in which drainage basin you are. Much larger differences that are easier to measure exist in other systems, between freshwater and seawater, for instance."

"Hmm, so you could determine whether an animal drank freshwater or seawater without measuring salt content, by just looking at the stable oxygen isotopes in the bones of that animal." I finish my beer and consider that someone could determine where the water in it came from by studying the isotopes in my body tissues. So if I got all my fluids from beer at the Del Rio here, they could determine that from a sample of my blood or my bones. Forensic bar science, so to speak. The thought is mildly disturbing, but I can see the scientific potential.

The conversation stays with me as we both leave Michigan, and as my research focuses more and more on fossil whales. Years later, with drawers well stocked with teeth of *Ambulocetus* and *Pakicetus,* I call her up.

"Lois, we're finding all these fossil whales in Pakistan. The pakicetids only come from freshwater rocks, the ambulocetids from coastal sediments. I think that these whales are making the transition from land to water right where I work in Pakistan."

"Yes, I have read your papers," Lois says matter-of-factly.

"Modern whales ingest seawater, and they had land ancestors that presumably drank freshwater. Could we analyze the bones and teeth of those fossil whales and determine what they were drinking, and determine what the time was when they switched from freshwater to seawater?"

"Sure. There are a number of conditions. You need an associated fauna so you can study the context, you need to know body size, you need modern analogues, you need to—"

Lois is about to go off on a complicated disclaimer, but I do not want to lose the momentum, so I interrupt. "I think that we can get all those things. How big a sample do you need?"

"Well, you need about five grams, and it would be good to have tooth enamel, dentin, and—"

"That does not mean anything to me. How big a piece of bone is that?"

"It depends on the thickness."

"How heavy is a fingernail clipping?"

"I don't know. I will need more than that."

My turn to get slightly miffed. I would like to move beyond distracting details. I want to know how many teeth need to be sacrificed for this, but Lois will not be drawn into vague analogies. It would be amazing if we could track the shift from freshwater drinking to seawater. To determine such an important evolutionary change—who could have thought that you could figure that out from fossils? Whales obviously would be unable to travel great distances across oceans if they needed a freshwater source, so the ability to drink seawater may have been a seminal moment allowing them to disperse across wide oceans.

Eventually, we decide that this is worth a try. I will take some enamel samples, and send them to her. She will chemically pry the oxygen atoms out of the enamel of the tooth, and lock them into a much larger molecule in the same ratio of ^{16}O and ^{18}O as was present in my teeth. Then she will fire those molecules through one of the arms of the mass spectrometer, and determine the proportion of abundance of the two molecules. That will allow us to see whether that ratio is closer to freshwater or seawater. To see whether the theory that we know of actually matches the real world, she will also run some enamel from modern oceanic whales and dolphins and compare that to the few species of dolphins that spend all of their time in rivers.

DRINKING AND PEEING

Lois sends me reams of background data on the method. Although excited, I am also worried. Is this really going to work? I am still skeptical in spite of my conversations with Lois. Can such fleeting behavior as drinking really be gleaned from these fossils?

In the meantime, I study what is known about modern marine animals and the drinking of seawater. For a thirsty animal, the problem with drinking seawater is that there is a lot of salt in it. In fact, it contains more salt than the blood and body fluids of a mammal do. As a result, if an animal drinks seawater to hydrate itself, it needs to take some of the salt out and excrete it, so that the saltiness of the new water matches that of its body fluids. In birds and crocodiles, there is a gland near the eye where salt is excreted. Mammals never have such a gland.

Land mammals lose salt when they sweat, but when they live in water, they can't sweat, since the process is driven by evaporation from the skin. The organs responsible for salt excretion in a cetacean, then, are the kidneys.

To take salt out of ingested water, the animal has to dissolve it in the urine it excretes by making that urine saltier. So the concentration ability of the kidneys is crucially important. Many small mammals, such as mice, can excrete highly concentrated urine and therefore can drink seawater.[1] Human kidneys cannot concentrate urine that strongly. In fact, for human kidneys to remove enough salt from seawater to match human body fluid, a lot of water is needed. That amount of needed water is greater than the seawater from which the salt is extracted. A human who drinks a batch of seawater will lose more water peeing out the salt in that batch than was gained drinking that batch. For human kidneys, the ocean is like a desert: there is no potable water.

Marine mammals cope with an absence of freshwater in different ways. In a rather nasty experiment, a sea lion named Dave was locked in a cage and only given seawater to drink, and his food was laced with salt pills.[2] Dave knew better than to drink the water, realizing that he would lose more of his precious body water excreting the salt. For more than a month, the animal did not drink at all, and seemed in reasonable physical health, until, mercifully, the experiment was stopped. Apparently, sea lions can withstand such dehydration. In contrast, if you hang a running hose into the ocean in Florida, manatees may swim up to drink from it. In spite of living in the ocean, they need a source of freshwater. At the other extreme, sea otters along the Pacific coast can drink seawater freely.[3] Cetaceans cannot concentrate urine to the levels where seawater drinking becomes an option.[4] Although they are known to ingest some seawater,[5] they get most of their water from their food, and use water very sparingly.

FOSSILIZED DRINKING BEHAVIOR

So, the isotope method may be able to answer the question of when cetaceans learned to cope with the absence of freshwater. Lois first runs the modern samples: pieces of teeth of some marine dolphins, a killer whale, and a sperm whale, as well river dolphin samples from the Amazon, Ganges, and Yangtze Rivers. To my delight, there is a consistent difference between marine and freshwater species,[6] and it matches what we predicted (figure 36).

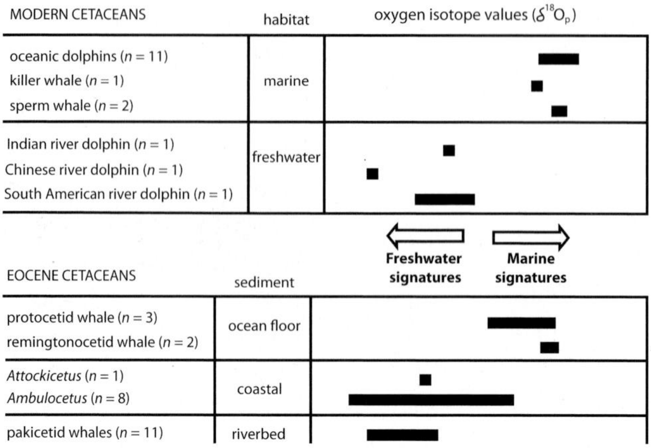

FIGURE 36. Oxygen isotope values for modern and Eocene cetaceans. Known habitats for the modern species indicate that isotope values can be used to determine whether they are freshwater or marine animals. Oceanic dolphins, killer whales, and sperm whales live in seawater, and their oxygen isotope values are high. River dolphins from different continents all have lower values. Thus, isotope values (indicated by isotope geochemists by δ, in which the ratio of the two isotopes is compared to a reference sample) can be used to identify water ingestion behavior for the fossil Eocene species. The isotope values for the fossils are consistent with evidence from the sedimentology of the rocks they are found in (ocean floor, coastal, or riverbed). This plot is based on data from Roe et al. (1998), and these results were confirmed and refined by more modern data from Clementz et al. (2006). Protocetid whales will be discussed in chapter 12.

Now the fossil work starts. My heart cringes every time that I have to break a piece off a fine fossil tooth to get an isotope sample. I worked so hard to get those teeth and not damage them, and now I am taking a screwdriver to their shiny enamel. I put the pieces in little vials and mail them off to Lois, who will grind them to powder. Lois sends me data and patiently explains what all the numbers mean.

"For the pakicetids, there is a clear freshwater signature," she says.

"Signature?"

"Signature means that the implication of the delta ^{18}O value is that it was freshwater."

I know about the delta bit—it is basically the ratio of the ^{18}O and ^{16}O. A lower delta value indicates more of the lighter isotope. Cool, but not surprising: they lived in freshwater, and that is what they were drinking, like any self-respecting land mammal. The Indian whales, such as the

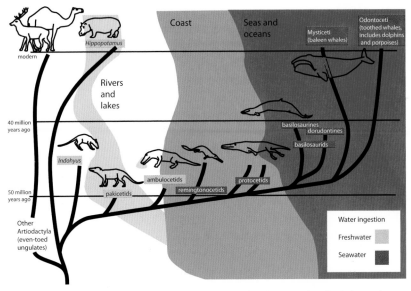

FIGURE 37. Branching diagram showing the relationship between fossil whales and artiodactyls, and the habitats in which they lived (shades of blue for water, white for land). Water ingestion behavior is indicated as the shades of gray of the boxes with names, and is based on isotope data. Protocetids will be discussed in chapter 14, and *Indohyus* is an even-toed ungulate that will also be discussed in chapter 14.

remingtonocetids, are on the marine end of the scale. Not just cool, but also surprising. Those whales lived in the sea near the coast, and their isotope values show that they are independent of freshwater altogether. This means that within a few million years of entering the ocean, whales did not require a freshwater source and could travel across oceans (figure 37). That is an interesting contrast with modern manatees. They originated around the same time as whales, but they still haven't figured out how to live on seawater only.

Of course, history matters here. Whales and manatees are derived from different land ancestors, and isotopes only show you what an animal drank, not what it would be able to drink if it had to. It is possible that the pakicetid body would be able to handle seawater, but since they lived in a freshwater ecosystem, they never needed to. Given that at least some modern artiodactyls can process seawater, the ability to handle life without a freshwater source may already have existed in the ancestors of cetaceans.

A few years later, Lois leaves science, and Mark Clementz takes over the isotope work. Mark is a generation younger than I, very dynamic and with a whole new array of sampling and isotope analysis techniques in his tool box. Most appreciated, from my perspective, is that now we only need tiny bits of enamel to analyze isotopes. I can barely see it when Mark drills into a whale tooth. Also, techniques have now improved so that we can answer much more sophisticated questions. For instance, Mark is able to use teeth that erupt earlier and later in the life of an animal and distinguish which of those teeth were formed (in the jaw) before the animal was born, when it was nursing, and when it became an independent feeder—all based on the differences in fractionation of isotopes at those stages.[7]

That level of detail will come in handy in studying *Ambulocetus*, whose isotope data are intriguing. Its isotope signatures are all over the place, but are mostly in the freshwater area. That is at odds with its coastal living environment, where seawater and brackish water abounded.[8] If the ability to handle seawater was not present in the ancestors of cetaceans, it is possible that *Ambulocetus* was living at the shore but had to swim up a river to drink freshwater in order to not overdose on salt. But other interpretations are equally possible. Maybe they lived in rivers as juveniles (when their teeth were formed), and moved out to the coast (where we find their fossils) later on. Or maybe they were like the alligators on Kiawah Island: they chose the freshwater habitats in an ecosystem dominated by marine habitats. If those alligators ever fossilize, they will be found among the seashells and shark teeth, in spite of their habitat. Maybe *Ambulocetus* did not drink at all, instead getting all its water from its prey—and *those* were freshwater fish or land mammals. Of course, we are first going to have to find milk teeth for *Ambulocetus*, and we haven't.

WALKING WITH *AMBULOCETUS*

With our isotope work on *Ambulocetus* in full swing, the whole habitat issue takes a turn into fiction. The BBC is making a series of documentaries about the evolution of mammals. They call it *Walking with Beasts*, and whales play a prominent role. *Ambulocetus* is seen swimming, walking, and hunting in one of the episodes. The makers do an excellent job trying to get the animal right. They send me version after version of little movie clips of the animal moving across the screen, first as a stick figure, later more and more realistic, and eventually with fur and a menacing glare. The makers take my comments seriously: the length of the snout gets fixed, and so does the flexibility of its spine. The outcome is stunning.

I am fascinated by its looks. It is as if a forty-eight-million-year-old film of the beast in the wild were discovered. The fascination comes to an abrupt end when they add the setting of *Ambulocetus'* appearance—they put it in a German fossil site, Messel. In the Eocene, Messel was a near-dead lake in a forest that belched toxic volcanic fumes. Most animals coming near it died of the fumes, and fossilization was common because the place was too toxic for scavengers to live and eat the carcasses. I argue with the makers of the show. *Ambulocetus* lived half a world away, on a desert coast, in waters that brimmed with life, not in a dead and deadly pond in the German forest. My complaints are acknowledged but rejected. To tell a good story, the whale needs to paddle happily through that toxic mud hole in pursuit of rat-sized critters on the forest floor. Don't believe everything you see on television.

CHAPTER 10

The Skeleton Puzzle

IF LOOKS COULD KILL

Locality 62, Punjab, Pakistan, 1999. Six of us are back at locality 62, the place where Robert West found the first *Pakicetus,* digging for more fossils of those elusive first whales. The vertical wall of hard reddish-purplish rock rises five feet out of the ground and the monsoons have washed it for us, exposing delicate, beautifully preserved fossils. The braincase that I saw some years ago is still there. The wall was originally not vertical, it was horizontal. The movements that formed the Himalayas pushed it up and superimposed a pattern of crisscrossing cracks, which make the wall look like it was built from carefully fitted, jagged stones. The fossils stand out in bright white and often run across a crack: after all, they were there before the rocks cracked. We take the wall down block by block, keeping track of adjacent blocks so as to not separate two parts of a fossil. We have a bit of a conveyer belt for fossils going. One of us numbers all the blocks when they are still in the wall; the next takes the individual blocks and brushes the dirt off them; then they are handed to me. I sit with a heavy hammer, chisel, and hand lens. I note where there are fossils and mark them, and someone else keeps the fieldbook up to date: "Five-pointed star; humerus, matches five-pointed star in Block 23." Two more people label and wrap blocks. The hunt goes well. We will be paying dearly in excess weight at the airport. If there are no fossils visible on the outside of a block, I smash it to see if there are some inside. If there are fossils, I trim the block, getting rid of unnecessary weight and thus saving money on shipping.

By far most of the teeth we find are whale teeth, pakicetids, but there are a few others. There is a tiny artiodactyl called *Khirtharia,* but we have only a few jaws of it. Artiodactyls, or even-toed hoofed mammals, include pigs and camels, goats and cows, hippos and giraffes, but *Khirtharia* is much smaller than those—the size of a raccoon, but totally unrelated to that carnivore. Teeth are actually not the most diagnostic part of an artiodactyl. All artiodactyls are characterized by the particular shape of a bone in the ankle, the astragalus. In all mammals, the astragalus is the bone on which the ankle pivots. To allow that, the bone has a hinge joint called the trochlea that articulates with the shin bone (tibia) above it. The other side of the astragalus is an area called the head. It faces the foot and has different shapes in different mammals. It is globular in most mammals, flat in horses, and has the shape of another trochlea in artiodactyls. This double-trochleated astragalus is very distinctive, characterizing all artiodactyls from the smallest mouse deer to the largest giraffe, including all the fossil ones. An astragalus for *Khirtharia* was among the bones collected by Davies and sent to Pilgrim in the British Museum, long before I was born, and before Dehm went and collected the first whale in Pakistan.

We also find lots of limb bones, and it is easy to identify those as tibias, femurs, or humeri. It is not so easy to figure out to which animal they belong. Given that most of the teeth are whales, most of the skeleton bones are probably also whales, but one cannot be sure. Size helps some—given the big size difference of the teeth, it is not possible to confuse a *Khirtharia* femur with that of a whale. Complicating this could be a shadowy species for which no teeth have been found yet, but only bones. That appears to be our problem here. There are a number of large double–trochleated astragali at this locality. They are clearly artiodactyl, based on their shape, but they are much bigger than *Khirtharia.* The species must have been pretty common, given that we have so many bones of it, but we have never found teeth of that artiodactyl. It is an enigma—but I do not worry about it. With the collection from this site growing, the problem will go away, and we will eventually have teeth and bones for all animals represented.

I muse about such matters as block 7 reaches me. It is so large that I have trouble lifting it, but parts of several fossil bones are immediately obvious on the outside. My hammer hits a corner of the block. Another bang, and I gasp. The rock breaks, and the crack exposes part of a braincase. It looks like the pakicetid braincase that Philip Gingerich found in 1981 (note 2 of chapter 1). That was a nice fossil, but the parts with the eyes, nose, and jaws were missing. As a result, we don't know what the face of *Pakicetus* looks like. The areas in the front of the braincase cannot be seen in this one either, but it is possible that it is in the adjacent block that is still in the wall.

I brush the new skull with water and a toothbrush, scraping dirt out of the cracks, so that it can dry and I can glue the weak points. This takes time. The others keep working, and a pile of blocks ready for smashing forms next to me. I ask the others to find the block that was adjacent to this one and wash it. Sure enough, the skull goes on into the next block. This could be exciting. Another bang, carefully placed, with adrenalin going to my head. The block breaks in two. My heart stops. The crack reveals the eye sockets of the whale, perched on top of the skull. The whale stares straight at me across forty-nine million years, as if the rock were muddy water and the whale were sizing me up as prey. I sit back and drop my hammer, and call the others to come and look. Here is the best skull of the first whale known to people—what a find. As I gently brush it off, I consider that the rest of its bones are at this locality too. It will just take time to extract them from these rocks.

HOW MANY BONES MAKE A SKELETON?

The critical question that we hoped to answer with fossils from H-GSP locality 62 was this: What are whales related to? More pakicetid fossils would answer that question, and this skull is an important part of the answer. For more than two decades, there was no controversy among paleontologists: whales are related to a long-extinct group called mesonychians (Mesonychia in Latin). It was an idea proposed by the brilliant and eccentric paleontologist Leigh Van Valen.[1] He observed that mesonychian teeth looked just like those of early whales. In both, a lower molar has a high trigonid with a single cusp and a low talonid with one cusp, very unusual for a mammal (figure 34). A lot of fossils are known for mesonychians—dentitions, skulls, and skeletons from North America, Europe, and Asia—and they lived at the right time to have given rise to whales.[2] Their dentition suggests a meat diet, and their body is vaguely wolf-like, but they are hoofed mammals: five toes per foot, with a tiny hoof at the end of each. However, the paleontological romance with the mesonychian-whale hypothesis encountered trouble from molecular biologists who found that, in terms of proteins and DNA, modern whales are very similar to modern artiodactyls. So similar, actually, that it appears that cetaceans should be included in even-toed ungulates: their closest relatives are hippos, and hippos are more closely related to cetaceans than to any other artiodactyl. That is called a sister-group relationship. Of course, mesonychians are extinct, and their proteins and DNA cannot be studied, and that leaves the possibility open that cetaceans and mesonychians were sister groups but that the two of them combined form a group that is the sister group of hippos. However, that did not sit well with the paleontologists either: the double-trochleated astragalus occurs in all artiodactyls, including hippos, but not in mesonychians, which seemed to exclude mesonychians from the artiodactyl group. In cetaceans, it is impossible to tell what the astragalus looked like since all modern and nearly all fossil cetaceans have lost their hind limbs. In basilosaurids, the ankle bones are fused into an unrecognizable lump, and in remingtonocetids, no ankle-bones are known. *Ambulocetus* was disappointing in this regard too: we found half of an astragalus, but not the part that would have solved the problem.

This is why a pakicetid skeleton is needed. It would provide a skeleton of a cetacean sufficiently primitive to allow us to make direct comparisons to artiodactyls and mesonychians. Ankles would be of particular importance to solve the artiodactyl-mesonychian riddle.

So, back in the United States, that is what we are going for. Ellen is extracting the bones from the blocks from locality 62 in the hope of finding enough of them to build a skeleton of a pakicetid. The trouble with the locality is that there are no single skeletons: the bones of lots of individuals and species are jumbled together here. Ellen has prepared drawers full of locality 62 bones, and there are lots of whales, given the teeth and skulls, but I cannot directly recognize which limb bones and back bones go with those teeth.

I open the drawers frequently, fitting humeri on radii and tibias on astragali. As I play with that unique jigsaw puzzle, missing pieces haunt me. I pull some of the most common bones out of the drawer. They belong to an animal that must be similar in size to the pakicetids. I put the bones on a table. The foot bones together make a well-proportioned foot, but it is not that of a whale, it is an artiodactyl instead: it has a double-trochleated astragalus. The two middle toes are similar in length, and much longer than the side toes. That is another artiodactyl feature: each foot has even numbers of equally sized toes (usually two long ones and two shorter ones, or just two toes of similar length). This foot belongs to a

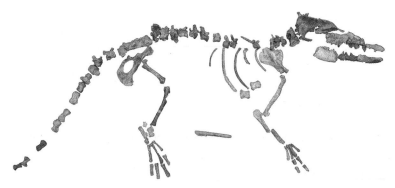

FIGURE 38. The skeleton of Eocene whale *Pakicetus,* put together from the bones of a number of different individuals, all washed together at Locality 62 in the Kala Chitta Hills of Pakistan, approximately forty-nine million years old. Study of stable isotopes confirmed that these all represent bones of this early whale species. The marker between the legs is 13.5 cm in length.

common locality 62 beast that really is an artiodactyl, so it is frustrating that I do not have any large artiodactyl teeth from here. Ellen takes more fossils out of the rock daily, but there are no large artiodactyl teeth.

The bones become an obsession. I leave them out on the table. The vertebral column; the shoulder, forelimb, hind limb. But there is no skull or teeth. I take one of the pakicetid skulls and put it at the front of the skeleton. It fits the first vertebra (the atlas) very well, and size-wise, it matches the skeleton (figure 38). It would solve the problem of the mystery artiodactyl: the mystery artiodactyl is actually a whale. Ellen walks in to show me a new bone that she just extracted from the rock. She sees what I did and blushes, which she does easily. The skeleton on the table is making a reckless statement about whale evolution: if that beast has a double-trochleated astragalus, Van Valen's great insight that whales are derived from mesonychians would be wrong. Disturbingly, it would mean that the teeth were lying to us—the detailed similarities between mesonychian teeth and pakicetid teeth would be convergences, not related to having a common ancestor.

Ellen and I ponder what to do next to see whether the fossil evidence supports the idea. To give this molecular biology–inspired idea a chance, we first have to study the relative abundance of fossils at locality 62. We count all the bones and teeth. Of the teeth that I can identify without doubt, 61 percent pertain to pakicetid whales. The bones are harder to count—there are so many of them, and all the different kinds have to be counted separately. After all the counting, the bones that are of the correct size to fit an artiodactyl with that mystery astragalus are more common than any other bones, just as the whale teeth are more common than any other animal's teeth. One would expect the most common teeth to belong to the same animal as the most common bones at a fossil site, so that supports the match between whale skull and artiodactyl skeleton.

Then, we look at the other animals known at locality 62. First is *Khirtharia*, the raccoon-sized artiodactyl, which makes up 14 percent of the identifiable dental fossils, the second-most common beast. We compare its teeth to those of artiodactyls from Messel, the toxic lake site in Germany, where entire skeletons are preserved, articulated as if their owners could jump up from the rock and run off. The mystery artiodactyl bones are much bigger than the bones of a Messel artiodactyl that has teeth the size of *Khirtharia*. Clearly, whales cannot be confused with *Khirtharia*. The whale-artiodactyl hypothesis passes another test.

About 11 percent of the jaws and teeth at locality 62 belong to an anthracobunid, the putative elephant-manatee relative, and it is the third-most common mammal. Their jaws are bigger than those of pakicetids, and there are several large bones at locality 62, much larger than the bones of the mystery artiodactyl. At a different place in the Kala Chitta Hills, we found a partial skeleton of an anthracobunid: teeth, skull, and bones, all of one individual. The bones are short and squat, and match the proportions of those at locality 62. They are very unlike the long and gracile bones of the mystery artiodactyl. Another roadblock eliminated.

I feel good about this, but it would be nice to confirm it with another line of evidence. Isotope geochemistry comes to the rescue. Stable isotopes of carbon in pakicetid teeth and jaws at locality 62 are very different from those of the teeth of the other mammals. Do they match the bones? I eagerly await the results of Lois's isotope study. They are a match! Pakicetid dental isotope signatures match those of the bones of the mystery artiodactyl, and they are different from the teeth of the other mammals. The conclusion now becomes inescapable. Ellen and I lay the skeleton out once more. It is the same skeleton, but now the discomfort is giving way to a sense of amazement, and of victory.

FINDING WHALES' SISTERS

Although finding and identifying the astragalus (figure 39) seems a seminal moment in our thinking of the relatives of whales, it will not be

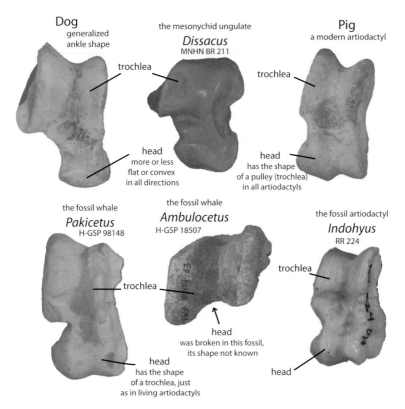

FIGURE 39. The astragalus is the bone on which the ankle pivots in mammals. The dog astragalus shows the primitive condition, where the head is more or less convex in all directions, while the top part, on which the ankle pivots, is a trochlea (pulley). In artiodactyls, the head also has the shape of a trochlea. Whales that still have hind limbs, such as *Pakicetus*, have an astragalus similar to artiodactyls. For *Ambulocetus*, that part of the bone was not found. *Indohyus* is a close relative of *Khirtharia*.

enough to convince the world. That will require an explicit consideration of all of the morphology of the cetaceans and all of their potential relatives: a cladistic analysis. In a cladistic analysis, all differences between animals are compiled in a table called a character matrix, and all of those differences are explicitly described. For instance, the shape of the astragalus is a character of relevance, and one could describe that character as having two states: "astragalar head has the ball-shape of a condyle" and "astragalar head has the pulley-shape of a trochlea." Numbers are then assigned to these states, usually zero and one (more if it is a complex character), and the computer maps those on different cladograms and calculates how many evolutionary changes would take place (figure 40). Our character matrix for the whale work takes its characters from our own work, but also from that of colleagues in the whale field such as Zhe-Xi Luo, Mark Uhen, Jonathan Geisler, and Maureen O'Leary.[3] The addition of a pakicetid skeleton to the matrix in a cladistics analysis showed indeed that mesonychians should be evicted from the extended family of cetaceans.[4]

Determining How Animals Are Related

Our character matrix has 105 characters that are columns of numbers, mostly zeros and ones. The twenty-nine species studied are the rows in the matrix. They include pakicetids and *Ambulocetus,* as well as artiodactyls from hippos to mouse deer, and several mesonychians.[4] The computer makes sense of the matrix by trying out possible combinations of proposed relationships and calculating how many evolutionary changes each would take. For instance, the computer will propose that *Ambulocetus* and pakicetids are sister groups, and that their next-closest relative is one of the artiodactyls, and this can be summarized in a cladogram (simplified version in figure 40, top). The computer then determines where on the cladogram each character would change given the particular relationship proposed in the cladogram and taking into account what the state of that character is in a group that is the most distant relative of all of them (the outgroup). For instance, we can plot the astragalar character on the top cladogram of figure 40, rooting it in primitive ungulates that have an astragalar head in the shape of a condyle. Hence, at the base of the cladogram, the astragalar character is in the zero state. Moving to the next branch on the cladogram, artiodactyls have a head that looks like a trochlea, so that means that an evolutionary change took place at that line segment from zero to one, as indicated by the short dash and the arrow between zero and one. Since *Pakicetus* is similar to artiodactyls, no change took place at the next branch, or any other branch.

To reason through this for multiple characters instead of one is too complicated for a human brain, but the computer does it by trying other hypotheses of relationships, as for instance in the second cladogram of figure 40, where mesonychians, not artiodactyls, are the sister group to *Ambulocetus* and *Pakicetus.* In this cladogram, the change leading to the astragalar head in the shape of a trochlea takes place on the branch between primitive ungulates and artiodactyls (change from zero to one), and then that character reverses to its original state (change from one to zero) at the branch between artiodactyls and mesonychians, to reappear once more (zero to one) at the branch between the mesonychians and the cetaceans. This evolutionary hypothesis takes three steps. If evolutionary change is rare, then the relationships suggested by this cladogram are less likely than those of the first cladogram, which only took one step.

But wait, we can maintain the relationships of the second cladogram and yet decrease the number of evolutionary steps. The third cladogram proposes that the re-emergence of a condyle-shape could occur on the line to mesonychians instead of in the common ancestor of mesonychians, pakicetids, and *Ambulocetus*. Even though the second and third have identical branching patterns, they make different statements about the evolution of the shape of the astragalus, and the third takes only two evolutionary steps. That is still more than the arrangement of the first cladogram, so the computer will point to the first cladogram as the most likely reflection of what happened in real life. It is easy to imagine that this gets more complicated if we have to try all possible branching patterns for twenty-nine species, and impossible to do by hand if there are 105 characters that do not evolve in unison and often point in conflicting directions. However, the computer can keep all of this straight and figure out the branching pattern that requires the fewest evolutionary changes. That is called the most "parsimonious" cladogram.

The field of systematics studies the relationships among animals by using the cladistics analyses explained in the sidebar. Esoteric as it seems to most laypeople, it is one of the most contentious areas of the study of whale origins, and some of the most argumentative scientists are systematists. Having said all of that, figuring out the relationships among the animals that you study is important for just about any other aspect of biology. The publication of the pakicetid skeleton with a cladistics analysis on all the whales[5] coincides with the publication of another Eocene whale skeleton from Pakistan by Philip Gingerich and colleagues,[6] and those papers seal the issue for most scientists: whales are related to artiodactyls. That does not mean that the fossil data are totally in agreement with the DNA data. The fossil data show that *some* artiodactyl (as opposed to a mesonychian) is the closest living relative of cetaceans, but it does not point to a particular artiodactyl as being in that position. A mountain of DNA data indicate that hippos are the closest living relatives of whales;[7] the fossils are just not that specific.

This bothers me. DNA data can never address the possibility that some fossil artiodactyl is even more closely related to whales than hippos are, because it is not possible to get DNA out of such old fossils. For

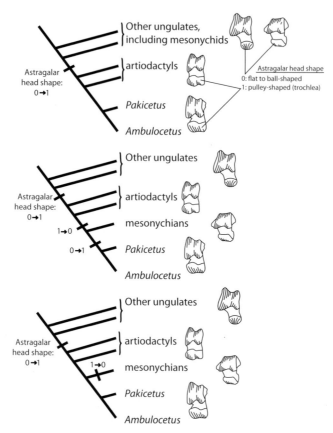

FIGURE 40. Three cladograms that show to which group of mammals cetaceans may be related. Most scientists support the top cladogram. Each of these cladograms has implications for how the astragalus evolved. A zero indicates that the astragalar head was convex; a one indicates that it was a trochlea; and an arrow indicates that an evolutionary change took place in one direction or the other.

now, I have to settle for less: the new evidence has routed the mesonychians in favor of artiodactyls as cetacean relatives. That is a big deal. Now we can focus on how pakicetids lived. In the future, I will be paying more attention to fossil artiodactyls as I think about whales, but for now, I indulge in a part of science for which I have had a weak spot ever since my first brush with whales, a long time ago: hearing.

CHAPTER I I

The River Whales

HEARING IN WHALES

The new pakicetid skulls can really help with learning about hearing. It was clear already that cetacean hearing changed when the ancestors of cetaceans went underwater. Land ears work poorly underwater, because sound in air differs from sound underwater. The fossils showed it too: that first pakicetid incus did not resemble modern whales or modern land mammals (figure 3); that thick involucrum must have done something to sound transmission (figure 2); and the mandibular foramen grew bigger over the course of the Eocene (figure 25).

In general, all the anatomical parts of the organ of hearing in whales can be found in land mammals too, but the shapes are different (figure 41). Land mammals have a canal in the side of the head that gives entry to sound: the external auditory meatus. It ends at the eardrum. Behind the eardrum are the three ossicles already mentioned in figure 3: malleus (hammer), incus (anvil), and stapes (stirrup). In most mammals, the ossicles are loosely suspended within an air-filled cavity, the middle ear cavity, which is protected by a protective bony shell, the tympanic bone in whales. The malleus looks like a club, its narrow handle firmly attached to the eardrum, and its wide part having a joint with the incus. As sounds make the eardrum vibrate, the malleus vibrates, and the vibrations are passed on to the incus. The incus has two arms, the crus breve and the crus longum. The former is anchored into the wall and

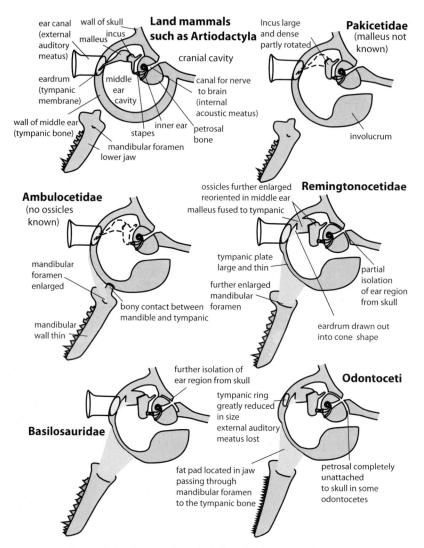

Land mammals such as Artiodactyla

ear canal (external auditory meatus)
wall of skull
malleus
incus
cranial cavity
eardrum (tympanic membrane)
middle ear cavity
canal for nerve to brain (internal acoustic meatus)
wall of middle ear (tympanic bone)
inner ear
stapes
petrosal bone
mandibular foramen
lower jaw

Pakicetidae
(malleus not known)

Incus large and dense partly rotated
involucrum

Ambulocetidae
(no ossicles known)

mandibular foramen enlarged
mandibular wall thin
bony contact between mandible and tympanic

Remingtonocetidae

ossicles further enlarged reoriented in middle ear
malleus fused to tympanic
tympanic plate large and thin
further enlarged mandibular foramen
partial isolation of ear region from skull
eardrum drawn out into cone shape

Basilosauridae

further isolation of ear region from skull
tympanic ring greatly reduced in size
external auditory meatus lost
fat pad located in jaw passing through mandibular foramen to the tympanic bone

Odontoceti

petrosal completely unattached to skull in some odontocetes

FIGURE 41. The ear in land mammals and whales. The diagram at the top left identifies all of the parts. Labels in other diagrams indicate which changes took place at each evolutionary step leading to modern whales. Dashed lines indicate bones not known for the group in question; their shape has been inferred from other groups.

helps in keeping the ossicles suspended and able to pivot. The crus longum has a joint with the stapes. As the incus pivots, the stapes is pushed in and pulled out of a small hole in yet another bone, the oval window in the petrosal bone. Behind the oval window is a cavity in the shape of a snail shell (the cochlea of the inner ear) that is filled with fluid. The pumping causes movements in the fluid, and that stimulates modified nerve cells that are arranged in a row along the length of the cochlea, passing the signal on to the brain.

In modern odontocetes (last diagram of figure 41), there is no open external auditory meatus; the duct is closed off by the tissues around it. The most sound-sensitive part of the face of a dolphin is actually the skin over the lower jaw, the mandible,[1] and sound travels from there through that large fat pad housed in the mandibular foramen of the lower jaw (figure 25). Sound constitutes vibrations that pass through a material, and these vibrations are passed on to the very thin part of the tympanic bone, the tympanic plate. Since it is made of bone, the tympanic plate has unique vibrational properties that are needed for the high-frequency sounds that odontocetes echolocate with. The eardrum is still present, but it is not a flat membrane. It looks like a folded-in umbrella. It may not have a function in hearing at all.[2] In whales also, the malleus is connected by bone to the edge of the tympanic plate; sounds are transmitted by the ear ossicles to the cochlea; and the latter works the same as in other mammals. The function of the involucrum is not well understood. It has been proposed that it is a counterweight during sound transmissions of the tympanic plate,[3] but the exact sound-transmission mechanism through the odontocete middle ear remains controversial, and sophisticated computer modeling of this area suggests that mechanisms may be different for different cetacean species and even at different frequencies.[4] The ossicles are much heavier in whales than in land mammals. That is strange—sound does not carry much energy, and it would be easier for faint sounds to make those ossicles vibrate if they were lighter. Possibly, the ossicles do not vibrate as one unit, but just parts of them vibrate, and maybe that process is helped by combining a thin vibrating process with the big inertial weight of the rest of the ossicle. That would especially be useful for high-frequency hearing—but this is all speculation. It is very difficult to study movements of the ossicles in a cetacean.

It would appear that many of the specializations of hearing in modern whales are actually for high-frequency echolocation. Odontocetes such as dolphins emit high-frequency sounds through specialized organs

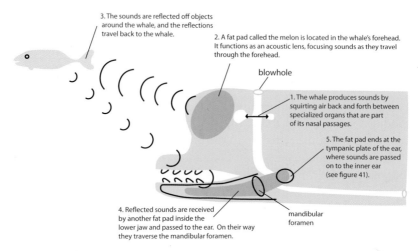

3. The sounds are reflected off objects around the whale, and the reflections travel back to the whale.

2. A fat pad called the melon is located in the whale's forehead. It functions as an acoustic lens, focusing sounds as they travel through the forehead.

blowhole

1. The whale produces sounds by squirting air back and forth between specialized organs that are part of its nasal passages.

5. The fat pad ends at the tympanic plate of the ear, where sounds are passed on to the inner ear (see figure 41).

4. Reflected sounds are received by another fat pad inside the lower jaw and passed to the ear. On their way they traverse the mandibular foramen.

mandibular foramen

FIGURE 42. The process of echolocation. The toothed whale (grey on the right) emits sound waves from its forehead. These reflect off the fish, and the reflections are received by the lower jaw and ear of the whale.

in their bulbous forehead and listen to the reflections of those sounds from potential prey with their sophisticated ears (figure 42). As a result, a blind dolphin can feed with little problem; a deaf dolphin will starve. In modern odontocetes, the stiffness of the tympanic plate and the heavy ossicles are adaptations for the perception of high frequencies, and not simply adaptations for underwater hearing.

Confusingly, the ear anatomy of baleen whales, mysticetes, is similar in many ways to that of toothed whales: the tympanic plate and heavy ossicles, and the shape of the tympanic membrane. But mysticetes are specialists at hearing low frequencies, not high ones. It is possible that the ancestors of mysticetes were high-frequency hearers, and that they retained some of the features of their ancestors but shifted others, to tune the ear to low frequencies (figure 43).

The ear is a wonderful organ to study for a paleontologist, because many of the important structures are bone and thus fossilize. For cetaceans, changes in the mandibular foramen, tympanic plate, and ossicles can all be studied in detail.[5] The closest Eocene ancestors of mysticetes and odontocetes are basilosaurids. They had a tympanic plate, a large mandibular foramen, and heavy ossicles of the shape of modern whales, and their tympanic membrane had the umbrella shape of their modern relatives. It is also clear that they were not echolocators, since they do not have the forehead organs needed to make echolocating sounds. It is likely

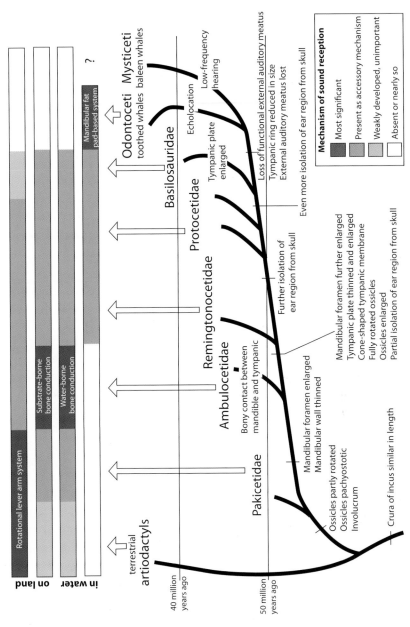

FIGURE 43. Cladogram showing evolution of features related to hearing. Bars on top summarize the evolution of sound-transmission mechanisms.

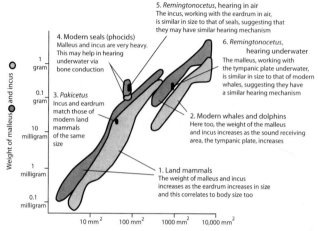

FIGURE 44. Mass of the malleus (hammer) and incus (anvil) of modern mammals and some fossil whales plotted against size of the sound-input area of the skull (eardrum in air, tympanic plate underwater in whales). *Remingtonocetus* may have had two sound-transmission mechanisms, one for airborne and one for waterborne sound. After Nummela et al. (2007).

that basilosaurids were specialized for high-frequency hearing, which is consistent with the idea that mysticetes had high-frequency ancestors.

All of these insights, inconsistencies, and opportunities dance through my head as I scrutinize the new pakicetid skulls. They have an involucrum like modern whales, but lack a large mandibular foramen and retain the external auditory meatus, which is also present in land mammals. The only ossicle we have, the incus, is heavier than that bone in land mammals, but lighter than whales, and looks different from, well, every other mammal incus (figure 44).

In air, pakicetids probably used the same sound-transmission mechanism as land mammals do: sounds make the eardrum vibrate and cause the ossicles to rattle. Underwater, it is likely that that system did not work very well. Instead, pakicetids may have heard by means of a sound-transmission mechanism called bone transmission, which does not allow for directional hearing. Humans experience bone transmission, for instance, when they are near loud, low-frequency sounds: the bass in a rock concert will send many of its vibrations through the floors and stands, and these reach the ear by passing through the person's body, not the air. Crocodiles lay their jaws on the ground and pick up the footsteps of their prey in that way,[6] and mole rats push their jaws against the walls of their tunnels to listen to sounds produced by animals in nearby tunnels.[7] Some forms of bone conduction are aided by the presence of heavy ossicles, and this may be the reason for the increased weight of pakicetid ossicles. From there, it may have been passed on to pakicetid descendants, including modern whales. Having said that, it is unlikely that pakicetids heard very well underwater, and they certainly could not distinguish where a bone-conducted sound came from.

Fossilized ears are also known for remingtonocetids. In this group (and also the protocetid whales, which will be discussed in chapter 12), the mandibular foramen is enlarged, the fat pad and tympanic plate are present, and the ossicles are large, similar to modern whales. However, these whales retain an external auditory meatus. These whales could still hear in air, but the heavy ossicles must have made efficient transmission of faint sounds difficult.[8] The mandibular fat pad was the sound transmitter underwater, just as in modern odontocetes. This new sound-receiving mechanism would make it possible for these Eocene whales to hear directionally underwater, as long as the pathways of bone conduction were switched off and could not interfere with the mandibular sound path. Bone conduction depends on a tight connection between the organ of hearing and the rest of the body, and such a connection is present in land mammals, as well as in pakicetids. But after pakicetids, that connection changes. The connection of the bones of the ear is looser in remingtonocetids than in pakicetids. In the former, a space occurs between the bones that hold the middle ear and cochlea (the tympanic and petrosal bones) and the rest of the skull. This space is larger in basilosaurids and later whales, and in modern dolphins and their relatives the space is so large that the ear bones tend to fall out of the skull when the soft tissues are removed. Moreover, in modern whales, that space is an air-filled cavity, similar to the sinuses in a person's forehead. That air is an acoustic insulator: it does not let bone-conducted sound pass to the ear. Undoubtedly, bone-conducted sound could cross to the ear in remingtonocetids, but the beginnings of the acoustic isolation that provides directional underwater hearing in modern whales are there too.

Not much is known about the ears of *Ambulocetus*. There is only one individual for the species for which the ears are preserved, and they are damaged by fossilization. However, it is clear that the species did have a partly enlarged mandibular foramen (figure 25) and a thin mandibular wall, both of which are involved in sound transmission through the jaw.[9] Most intriguing about *Ambulocetus* is that the jaw joint is expanded in such a way that the mandibular condyle (the part of the lower jaw that makes that joint) is in direct bony contact with the tympanic bone. That direct connection could also be a path for sound from jaw to ear, as it also occurs in mole rats. *Ambulocetus* may have been an early experiment to involve the lower jaw in sound transmission—far from perfect, but better than what pakicetids had—but if so, it was then quickly discarded in the evolutionary process with remingtonocetids.

Taken together, the ear story is intricate and exciting. Modern whales have ears that are relatively similar, well adapted for underwater hearing. The early whales show that hearing gradually changed and that there was an experimental phase, where the sound-transmission mechanism initially built for hearing in air was modified to allow bone-conducted hearing, an imperfect system, before a new sound-transmission mechanism evolved that was only perfected in early odontocetes. After that, the original land-mammal system was lost.

PAKICETID WHALES

The ears of pakicetids already suggest that they spent time in water; so if, in *Jurassic Park* fashion, we could bring one back and put it in a zoo, we had better keep that in mind (figure 45). On land, visitors would think a pakicetid was a wolf with a long nose and an oddly long and powerful tail (figure 46). Differently from wolves, though, we would watch them in the underwater viewing area, since they would spend much of their time wading in the water, spying over the water-line for unsuspecting and thirsty prey.

These earliest of whales all lived in a geographically small area,[10] in what is now northern Pakistan and western India (figure 22), around forty-nine million years ago. Just three genera are known: wolf-sized *Pakicetus* and *Nalacetus*, and fox-sized *Ichthyolestes*. *Himalayacetus* from India was also described as a pakicetid, but is more likely to be an ambulocetid. Locality 62 in the Kala Chitta Hills has produced more pakicetids than all other localities combined, but the site is a big jumble of the bones of many individuals; there never has been an associated skeleton of a single individual, so the reconstructions are composites (figure 38). *Ichthyolestes*'s small size helps in distinguishing its bones from those of the larger pakicetids. *Pakicetus* and *Nalacetus* teeth and tympanic bones are different in shape, but their limb bones are difficult to distinguish.[11]

FIGURE 45. Life reconstruction of *Pakicetus*, the first known whale. It is at the base of the cetacean radiation and lived forty-nine million years ago in what is now Pakistan. Externally very different from modern whales, dolphins, and porpoises, it was an amphibious wader that lived in shallow streams.

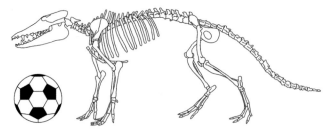

FIGURE 46. The skeleton of the Eocene whale *Pakicetus*. The soccer ball is 22 cm (8.5 inches) in diameter.

Feeding and Diet. A lot has been learned about pakicetid feeding in recent years, but many questions remain. Stable-isotope studies show that they drank freshwater and were flesh eaters,[12] and they have sturdy high-pointed front teeth, as is common in predators that grasp struggling prey. The premolars are triangular, and upper and lower premolars lock together into a zigzag pattern that cuts into the flesh of the unlucky victim (figure 47). Pakicetids have the same number of teeth as other basal placental mammals: 3.1.4.3 in both upper and lower jaw. The lower molars have a low and a high part (talonid and trigonid), and the upper molars have three large cusps (figure 34). Crushing basins and crests on the molars are reduced, and they lack the cutting edges that are found in carnivores; instead, their wear pattern is similar to that in other Eocene whales: steep wear facets that indicate that pakicetids chewed their food in very unusual ways.[13] That wear pattern does not occur in any modern mammal, and it is difficult to make sense of it.

In general, the amount of tooth wear in an animal depends on the kinds of food it eats, its age, and the way it uses its teeth.[14] From the position of wear facets on the teeth, one can determine how teeth rubbed along each other and how they interacted with food. Some wear, near the tips of the cusps, is caused by tooth–food–tooth contact before the upper and lower teeth contact each other as the jaw closes. This type of wear is called abrasion (figure 48). During chewing, abrasion is the first wear to occur. After that, cusps from opposing upper and lower molars slide past each other and cause a type of wear called attrition, resulting from tooth–tooth contact. There are two phases to this attritional movement. During phase I, teeth are coming into full contact as lower teeth shear along the upper teeth, moving up and somewhat toward the side of the tongue (lingually). Phase I ends when the upper and lower teeth come into full interlocking contact. In phase II, lingual movement con-

(crowns for three upper and lower incisors not preserved)

Upper teeth in occlusal view

One canine Four premolars (crown of first one lost) Three molars

Upper and lower teeth in side view (labial)

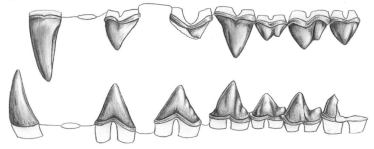

Lower teeth in occlusal view

One canine Four premolars (crown for first one not preserved) Three molars (last one damaged)

FIGURE 47. The dentition of the left upper and lower jaw of *Pakicetus*. Tooth crowns are known for all teeth shown. After L. N Cooper, J. G. M. Thewissen, and S. T. Hussain, "New Middle Eocene Archaeocetes (Cetacea:Mammalia) from the Kuldana Formation of Northern Pakistan," *Journal of Vertebrate Paleontology* 29 (2009): 1289–98.

tinues as the lower teeth slide further toward the tongue, but now the jaw opens slightly. At the end of phase II, upper and lower teeth lose contact, the jaw opens further and the cycle is repeated.

This precise interlocking of the teeth occurs only in mammals, and attritional wear facets are a characteristic of them. However, modern cetaceans are an exception to that mammal rule. They do not chew and do not occlude their teeth very precisely. There are hardly ever attritional facets in a modern odontocete. Most tooth wear in odontocetes is caused by contact with food: abrasion. This kind of tooth wear can be spectacular. Some killer whales wear their teeth down to flat stubs. It has been shown that those individuals suck in water with their prey and that they eat mostly small fish and occasional seals and large fish.

Right lower molars

The artiodactyl *Indohyus*

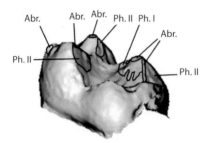

The cetacean *Pakicetus*

FIGURE 48. Three-dimensional reconstructions based on laser scans of lower molars of an archaic artiodactyl (*Indohyus*, discussed in chapter 14) and the ancient whale *Pakicetus*, showing the tooth crown morphology. Artiodactyl lower molars are characterized by three types of wear: abrasion (Abr.), phase I attrition (Ph. I), and phase II attrition (Ph. II). *Pakicetus* and other whales have teeth with simpler crowns and show nearly exclusively phase I wear. Compare with figure 34 to see how whale molars changed in evolution.

Counterintuitively, killer whales that specialize on feeding on large whales have barely any abrasional tooth wear.[15] Wear is equally impressive in beluga whales, another suction feeder mostly feeding on squid and bottom fish. When beluga teeth erupt, they are sharp prongs as in other odontocetes, but then the teeth quickly wear down to be nearly flat stubs (figure 49). This kind of sloppy abrasional wear is very different from the earliest whales.

Basal members of the artiodactyls are the closest land relatives of cetaceans, and they have unspecialized teeth that display all three types of tooth wear—abrasion, phase I attrition, and phase II attrition—in comparable amounts (figure 50). Tooth wear in early whales is extremely specialized. There is no phase II attrition and barely any abrasion. Phase I attritional wear dominates these teeth. It is not clear what this means. Was their prey special and in need of unusual ways of processing, or was this just the way that pakicetid ancestors chewed, without there being anything particularly good about chewing in this way? The early families of whales lived in a variety of environments, from freshwater to oceanic, but their wear patterns are similar, and the particular diet or food-processing mode was ubiquitous, regardless of environment. To understand what went on, we need to know exactly what live prey the early whales ate, and what whales' ancestors ate. We might be able to

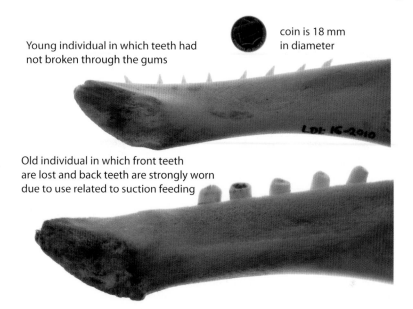

Young individual in which teeth had not broken through the gums

coin is 18 mm in diameter

Old individual in which front teeth are lost and back teeth are strongly worn due to use related to suction feeding

FIGURE 49. Lower jaws of a young and old beluga whale, with a penny for scale. In life, the teeth in the young individual had not erupted from the gums.

track diet by doing more in-depth isotope work. As for the ancestors, we need to sort through artiodactyls, in particular those from the time and place where the early whales lived: Asia in the Eocene. Artiodactyls are clearly critical to solving this puzzle.

Sense Organs. Clues regarding prey also come from the position of the eyes. In *Pakicetus,* they are close together and raised above the rest of the skull near the midline, and they face up, dorsal (figure 51). This differs from *Ambulocetus,* remingtonocetids, and basilosaurids (figure 52). The pakicetid position occurs in animals that live underwater but that watch what goes on above the water-line. Crocodiles, for instance, may sneak up on their prey with eyes and nose emerged but body and head hidden underwater. In hippos, the eyes are also elevated above the skull, enabling them to stay submerged while looking out above the water. It is likely that pakicetids lay in wait, hunting animals that came close to the water's edge. As discussed, the bone-conducted sound of the footsteps of prey may have been an important sensory cue.

The unusual position of the eyes affects the other sense organs. The nose and the nerves going from it to the brain are located between the

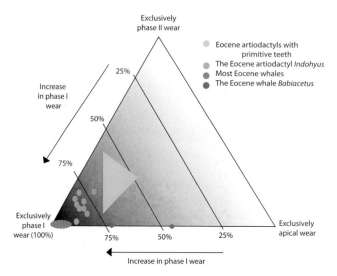

FIGURE 50. Diagram summarizing tooth wear on lower molars in ancient artiodactyls and whales. Surface areas of apical abrasion, phase I wear, and phase II wear are measured, and then recalculated as a percentage of total surface-area wear. These three kinds of wear are then plotted on axes that make up the three sides of this triangle, with the corners representing teeth with exclusively one kind of wear. In most Eocene whales (red oval), phase I wear dominates on the teeth, but most basal artiodactyls are closer to the center of the triangle (yellow triangle), indicating that they had all three kinds of tooth wear. Redrawn from Thewissen et al. (2011).

eyes and their nerves. In animals with large eyes that are close together, the structures related to vision seem to encroach on the space for olfaction. This is the case in humans: the nerves to the nose are moved to the area above the eyes, and they are small. This may be part of the reason why humans have excellent vision but a poor sense of smell. The same is true in pakicetids: the closely set eyes make the interorbital region (the area between the eyes) very narrow. For the fossil collector, this has the unfortunate consequence of creating a zone of weakness where most pakicetid skulls break during fossilization; and for the animal it had the consequence that the nerves coming from the nose that carry information about smells must be small as they pass through this narrow passage. The sense of smell of these first whales was limited. For reasons that are not clear, the interorbital region is not just narrow but also long, and as a result, the olfactory nerves and the bony tracks that they reside in are long. That feature is present in all early whales, and

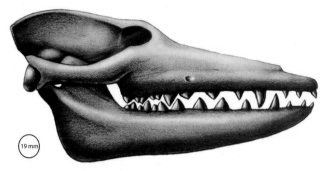

FIGURE 51. Skull of *Pakicetus attocki,* the most archaic whale, known from Pakistan. The circle is the size of a penny, 19 mm in diameter. Reconstruction based on H-GSP 18467 (braincase and orbit), 18470 (maxilla), 96231 (premaxilla), 30306 (maxilla), 1694 (mandible), and 92106 (tip of mandible).

can be easily seen in *Remingtonocetus* ("tract for olfactory nerve" in figure 35).

Pakicetids have a long snout,[16] but not nearly as long as in ambulocetids or remingtonocetids. The nose opening was near the tip of the snout, and bone in this area is perforated by many small holes through which, probably, nerves traveled. Nerves in this area usually relay information from the snout and whiskers back to the brain, and it is likely that pakicetids had a sensitive snout with many whiskers. Modern seals use their whiskers to detect vibrations in the water,[17] and it is possible that pakicetids did the same.

Walking and Swimming. The position of the orbits is not the only feature that suggests an amphibious lifestyle for pakicetids. The bones of the skeleton also indicate it. Limb bones of mammals usually have a large marrow cavity, surrounded by bone. The bone here has the shape of a cylinder; its outside is massive and is called the cortical layer. It is thinner in animals that need to be light, like bats, and thicker in those that need strong bones, like buffalo. In aquatic animals, the bones can be a source of ballast, allowing the animal to stay down and counteract buoyancy, so their cortical layers are often extra-thick. This is true for hippos and sirenians, for instance, and is called osteosclerosis (discussed before in chapters 2 and 3). Osteosclerosis does not occur in aquatic mammals, such as dolphins, for whom speed is important, because the weight would slow them down. Unlike most modern whales, pakicetids are osteosclerotic—their cortical

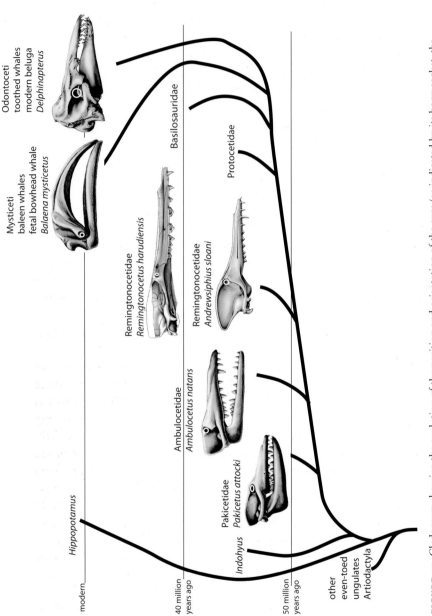

FIGURE 52. Cladogram showing the evolution of the position and orientation of the eyes (as indicated by its bony socket, the orbit) in some ancient and modern whales. White oval indicates position of the eye. Drawings are not to scale and thus not a good indicator of the size of the eye.

layer is extremely thick. The osteosclerosis of the limb bones suggests that pakicetids spent time in the water, but were not fast swimmers.[18]

So, how much did they move? On land, they could certainly walk. Their body proportions were similar to those of a wolf, but, given that the bones were so heavy, their locomotion was probably lumbering and slow. Just like land artiodactyls, the back was relatively immobile—the vertebrae of the lower back had interlocking joints that limited movement—whereas the joints of the legs allowed a lot of mobility in the front-to-back direction, and less in the side-to-side direction.[19] They had five toes on the hand, and four on the foot, with no indication that there was webbing. All fingers ended in a small hoof, betraying their ancestry as ungulates, but when they walked they were not up on their hooves, but instead had the entire finger touch the surface, like a dog, a pattern called digitigrady. The osteosclerosis of the limbs would have prevented fast swimming. Two features reveal a bit more about aquatic locomotion in pakicetids: the pelvis and tail. The pelvis of most four-footed mammals has a long part in the front, the ilium, and a shorter part in the back, the ischium. Those length relations are reversed in pakicetids: the ischium is proportionally long and has a large expanse for the attachment of the hamstring muscles. Hamstring muscles are large in animals that kick back their legs, such as seals. That may indicate that pakicetids did some swimming.

In addition, pakicetids have relatively large tail vertebrae. They are not as large as in *Kutchicetus,* and this is a tricky subject to study. Since there are no associated skeletons of pakicetids, it is not known how many tail vertebrae they had, and it is thus impossible to know exactly how long the tail was. Many tail vertebrae were found at locality 62, so it is likely that the tail was long. Furthermore, the number of tail vertebrae in the artiodactyls that were relatives of whales (*Messelobunodon,* twenty-four) and in cetaceans slightly younger than pakicetids (*Maiacetus,* twenty-one) are similar, so it is reasonable to assume that pakicetids had slightly more than twenty vertebrae too. From the shape of the fossil vertebrae that we found, we know that the tail was muscular. Hind limbs and tail may have been used to give the animal a burst of speed at the moment of attacking its prey, but it is unlikely that they were sustained swimmers.

Habitat and Ecology. When we first figured out which bones from locality 62 belonged to pakicetids, the most impressive thing about them was how gracile the limb bones were. They do not look like the stocky limbs of their closest relatives, other Eocene whales, but instead resemble those of more distant relatives, the running artiodactyls. As we studied the bones in more detail, it became apparent that a comparison with tapirs was more appropriate. Tapirs live in forests, but do like to be in the water when they have access to it: to cool off, eat water plants, and take refuge. Pakicetids were able to get around on land and in water too, but probably spent most of their time in freshwater ponds and rivers. Their anatomy reveals more aquatic adaptations than that of a tapir skeleton, and, just like tapirs, they were probably mostly waders.

All pakicetid fossils from Pakistan have been found in rocks that formed in shallow ponds, in a dry climate with occasional flash floods, as discussed in chapter 3. It is likely that these cetaceans lived like crocodiles, hunting by sitting still in the water, waiting for unsuspecting land animals to come and drink, or attempting to catch fish in the shallows. Pakicetids are rare at other localities in the Kala Chitta Hills. Those localities do have fossils of the land mammals that lived here, such as the small artiodactyl *Khirtharia,* brontotheres (cow-sized and rhino-like animals), small carnivores, and anthracobunids. All of them could be potential prey for pakicetids, as is suggested by stable-isotope evidence.[20]

SEPTEMBER 11, 2001

We published[21] our *Pakicetus* skeletons from Pakistan less than two weeks after the terror attacks of September 11, 2001. Soon thereafter, the eyes of the world focused on Pakistan and Afghanistan. It became more difficult to travel by plane in general, and specifically to work in Pakistan, and the last time I went there was 2002. Many people in the West began seeing that country as failed and lawless and inhabited by savages. Pakistan has failed in some regards, and there are lawless areas ruled by gangs of criminal bullies. However, they are not representative of all of Pakistan. In fact, some of the greatest acts of kindness and unselfishness bestowed on me have been by Pakistanis who had nothing to gain from doing so and could easily have gotten themselves in trouble for helping me.

One that I remember vividly occurred when Ellen and I flew from Cleveland to John F. Kennedy airport in New York on an American carrier and then on to Islamabad on Pakistan International Airlines (PIA). Our plane to JFK was very late, and the luggage was not checked through. We changed terminals with carry-ons and two suitcases of gear each. As we walked up to the check-in counter, groaning and sweating under our heavy load, four Pakistani-looking PIA officials were chatting, but the lights at the counter had been turned off. I asked the man behind the counter about the flight, and he said we were too late, pointing at the computer system—it was shut down—and he shrugged his shoulders apologetically. My face must have dropped, but then an older PIA man stepped in, and said, in strongly accented English: "Can you carry all that baggage through security?"

I said yes. He said something in a language that I did not understand to one of the others, who hurried off, and then something to the PIA woman, who talked into her walkie-talkie. I could just understand her Urdu words for "two" and "four."

"Good, follow this lady," he said.

She rushed us through security, down the terminal, to the gate. Our four suitcases were taken from us in the jetway; they closed the plane's door right after we entered, and the plane left with a slight delay. Traditional Pakistani hospitality and generosity—thank you, PIA.

CHAPTER 12

Whales Conquer the World

A MOLECULAR SINE

Tokyo, Japan, February, 2000. I think about the relatives of whales as
I travel on a metro train to visit the laboratory of Professor Norihiro
Okada. He goes by the nickname Nori, which is also the Japanese word
for a much-eaten kind of seaweed, as he points out with a broad grin.
Nori is not a paleontologist but a molecular biologist. The molecular
similarities between whales and hippos are piling up as more genes are
studied, and Nori is a central person investigating this. In his lab, dozens
of busy young people produce reams of DNA data. The DNA molecule
is like a string with four types of beads, the nucleic acids, and Nori's lab
spends its time determining the order in which the beads occur. Animals
that are more closely related will have more similar bead-sequences
than those that are more distantly related because there was less time
for the beads to change (mutations to occur) and the sequence of strings
to diverge. Nori's lab studies a special kind of DNA: short interspersed
nuclear elements, or SINEs.

I meet Nori in a tiny office which he shares with two secretaries. Nori
wants all his space to go to production—his labs—he does not want a
big private office. He sits on his mini-desk, barely large enough for a
computer, and I am on a tiny couch, unable to stretch my legs because
the tiny coffee table is too close. There is constant Japanese chatter from
the secretaries behind the bookshelf that partly divides the room, and

they bring us green tea. I do not like green tea, and this kind reminds me of water that was used to boil spinach in. I add a lot of sugar to mask the flavor.

Unlike many of his countrymen, Nori's English is well pronounced and articulated, although he often has to stop to think of a word, and plurals and articles are rare.

"SINE method is very useful method. Insertion of SINE is unique event."

"So, tell me how SINEs end up in the genome of an animal—how are they inserted?"

"Ahh… SINE are retroposon. SINE is very common, in humans 11 percent of genome is SINE." I think the gasp was at my ignorance, but he appears willing to educate this humble fossil guy nonetheless. It emboldens me, and I seize the opportunity to learn. His answer did not answer my question in a way I understand, so I try again.

"What is a retroposon?"

"Ahh… retroposon was inserted in host genome, maybe by help of virus element called a LINE."

I do not know what a LINE is, but this still clears matters up some. I know about viruses. Initially, a fragment of genetic material may have been part of the genome of a virus. When a virus infects the cell of a mammal, it injects its genetic material into the host cell; there it is incorporated into the DNA of the mammal. The mammal cell keeps on dividing, and, inadvertently, also duplicates the virus DNA. That allows the virus to take over the reproductive equipment of its host cell and make a new virus. Some of those inserted parts are SINEs—pieces of DNA that were initially part of the virus and not part of the DNA of the mammals' ancestor.

"So the host cell cannot recognize these SINEs and has no way to get rid of them?"

"There is no known mechanism to delete SINE." I can see that that would be useful to determine ancestry. If some little ribbon of DNA, a SINE, is inserted in the ancestor of an animal and can never be deleted, it would be present in all of its descendants, and thus would be a great marker to determine relationship between its descendants, since animals descended from a different ancestor will not have that ribbon of DNA.

"Is it not possible that a SINE is inserted in the genome of two different mammals independently? How do you know that a SINE that you find in the DNA of two animals is not the result of two separate insertion events in their ancestors?"

"SINE insertion in genome is not site-specific. We determine flanking sequences. These have to be same in case they are part of same insertion event. Probability that insertion of SINE is in same region of different hosts is close to zero."

This makes sense. If some viral SINE DNA is inserted into the genome of a host, it could end up anywhere in that genome. The chance that the same SINE is inserted into the same stretch of DNA of a host independently in two species is very low. I ponder the implications. If what he says is true, then this is a great way to figure out relationships. The SINE sequence can be inserted anywhere among the millions of genes of the host, and does not affect the function of the cell it is in. If there is no known mechanism that allows cells to cut out these inserted SINEs, and if they are neither harmful nor beneficial to the host, selection does not act on them. They just sit there and are copied, generation after generation.

That gives molecular biologists a great tool to figure out who is related to whom. As it turns out now, hippos have a SINE in common with whales, and it is found in the same place in the genome of the two groups.[1] SINEs in common between hippos and cetaceans imply that they were inserted into the genome of the common ancestor of those animals, but not into an earlier ancestor that was also ancestral to cows and pigs, since they do not have that SINE. That implies that hippos and whales are more closely related to each other than either is to cows and pigs.

Working with Nori makes me accept that the molecular evidence linking hippos and whales overwhelms dissenting fossil evidence to the contrary. But it also makes me see more clearly what role the fossils still have to play. The biggest problem with thinking of hippos as close relatives of whales is that the oldest hippos are only about twenty million years old,[2] nearly thirty million years younger than the oldest whales, and that, body-wise, the similarities are very limited. The long ghost lineage of hippos, between forty-nine and twenty million years ago, implies to me that the ancestors of hippos were so unlike modern hippos that we do not recognize them, so we really do not know what that last common ancestor looked like. Personally, I feel that we need to look for something that lived around the time of the earliest whales, close to the common ancestor of whales and hippos. But not in Kutch—the rocks there are marine, and they are too young. I have to explore other places. The older rocks in Pakistan are now unsafe to go to—maybe places where I have never been in the Indian Himalayas. I make a mental note to that effect.

THE BLACK WHALE

It will take time to start elsewhere, and Kutch is still producing interesting fossils. I consider how sad it will be to stop working in Kutch as we drive through its desert to the locality Dhedidi North. Forty-one million years ago, Dhedidi North was a lagoon that was slowly drying out (figure 30). There are some cool fossils here—a snake skull larger than my hand and a crocodile snout longer than my leg. Based on those, it must have been a scary place to walk around in back then. All these beasts perished the same way: their death came as they were trapped in the hot mud that slowly dried out and they were baked inside as their lagoon dried up. Many of them became ugly fossils, because gypsum dissolved in the water precipitated upon evaporation and formed a crust around the bones and teeth. Crystals also grew in the little cavities inside the bones and cracked and split the bones open.

The place is more pleasant in modern times. We park our car on the high yellow ledge, the Fulra Formation (Fulra Limestone, figure 28), and walk down into the gypsified mudstones of the Harudi Formation on one of the many trails made by roaming cattle. The nearby village of Dhedidi is traditionally inhabited by milkmen. Milkmen here do not buy and resell their milk; instead, they have their own herds of bovine producers, and those graze on the sparse grass around me. In the morning, the milkmen ride off on bicycles or motorcycles, metal jugs dangling from their handle bars and luggage racks, to peddle milk door to door in the villages.

Much of the fossil bone here is black, and the colors of the rock vary from ochre red to yellow and brown, plus bright-white gypsum, distributed in no particular pattern. The gypsum crystals grow in regular shapes that are suggestive of the regular shapes that fossils have, and I pick up many pieces of presumed fossils that on closer inspection disappoint. As I climb down a low hill, a row of five shapes the size of oranges attracts my eye. When I kneel down, they turn out to be vertebrae, arranged just as they were when they were still in the animal, millions of years ago. Usually this would be very exciting, suggestive that much more of this animal was buried, but these particular ones do not enthrall me because they are in terrible shape. Gypsum surrounds them on all sides, and they are weathered into jagged shapes. It is as if someone with a bone-cutting knife randomly hacked at them. I can just recognize them as rib-bearing thoracic vertebrae, and, sure enough, scattered around them are pieces of ribs and other vertebrae. The ribs are not pachyostotic. This was a fossil whale. I gather the loose-lying fragments in piles and then dig in the place where the quintuplets protrude from the hill.

On one side, there clearly is nothing. Weathering has excavated it for me, decades or centuries ago, and pulverized what it found. On the other side, the sediment is not eroded, and I run almost immediately into another vertebra as I start to dig. This vertebra is much better preserved—black and with just a bit of gypsum, but also with several of its processes intact. The fossils are fragile, and it requires a number of cycles of brushing out dirt, gluing cracks, letting glue dry, and exposing more of the fossil. A second vertebra is located immediately behind it, and this one is articulated with the first. Both are partly gypsum-encrusted, and that slows down the process of excavation. The two collectors with me have noticed that I have not moved for a while, an indication to them that I found something. They come over to help. We remove the overburden and excavate further, finding more vertebrae. The row of vertebrae snakes around into the muddy knoll I'm sitting on, and I suddenly realize that this is indeed the pattern expected from an animal whose body dried out when the lagoon dried up: the ligaments shrink as the body dries out, curving the vertebral column into a backbend. Could there be an entire skeleton?

Thinking again, I reject the idea. This place was clearly disturbed. I have thoracic, lumbar, and caudal vertebrae, but no pelvis or hind limbs. Much of this guy is gone. It is unlikely that erosion did that, because the places where those bones would have been are undisturbed. Instead, I suspect there were scavengers before the animal was buried. The dig continues and, to my surprise, we find four fused vertebrae, the sacrum. But the bone is not near the lumbar vertebrae as it would have been in the living animal. Furthermore, it is smaller than the lumbar vertebrae, and rust-red in color, not black. The edges are oddly and smoothly worn, in a different way from the sharp breaks in the vertebrae, as if someone knocked all the edges off before it was buried. It puzzles me as I consider the option that this little red sacrum was in the belly of the big black whale and is evidence of one whale eating another. At this point, that is just speculation.

We excavate the hill with the pointed backs of our hammers, moving along the entire edge where the skeleton is visible. On the side away from the vertebrae, I encounter a flange of thick, black bone, the shape and size of the visor on a baseball cap. We excavate around it carefully, but I cannot figure out which part of the whale this is. Impatiently, I pull at it, and it comes loose. It shows a break—it was attached to something bigger, and I just broke it. That annoys me. We continue to excavate the bigger thing. That takes a lot of time. It is deeply buried and hard to access. We lie on our bellies in the muddy soil, the hot sun burning our necks. Because the sediment is wet, our glue dries slowly, requiring patience that I am unable to summon and which does nothing to improve my mood. A larger black bone is finally exposed, bigger than I expected, and with an unusual triangular shape. I suspect that this fossil is unrelated to the whale vertebrae and that it is a skull of one of the giant catfish that used to live here. Those are quite common, but they are mostly ugly and gypsified fossils, and not very interesting paleontologically. I move faster to get through this, but the bone does not cooperate. It makes me more impatient, so I just tug on the fossil. The big black fossil suddenly comes loose. It is the size of a soccer ball, and I fall backwards holding it in my hands. It is the braincase of a large whale—the visor flange was the flange in the back of the skull where the neck muscles attach. It is now also crystal-clear that I messed up: in my impatience, I have damaged the fossil whale skull. Instead of pulling at pieces and breaking them off, I should have slowed down, excavated the soil around the fossil, and not pulled on the fossil until it was visible in its entirety.

I sit down, mad at myself, mad at the black skull, and mad at most of the rest of the world. After a drink of water and some cookies, my head clears, and a new plan is hatched: limit the damage, be patient, and excavate carefully. The braincase is well preserved, and I can see that more of the specimen is still in the hill. Everything slows down now, but we do it right, we remove overburden carefully, brush surfaces clean, and let them dry before we glue them and continue with the excavation. Eventually, much of a skull emerges, though somewhat gypsum-encrusted, and deformed by burial forces forty-two million years ago. The snout is complete, but it is waterlogged and the bone is soft as dust. I excavate small parts of it with a dental scraper, let them dry, and then harden them. Eventually the snout comes out, with lots of teeth in place. We pick up many small bone pieces, and put them in bags. At home, we wash all of them, and attempt to fit the smaller pieces to the big pieces. All the big pieces fit together. This is a very different whale from the remingtonocetids. Its big eyes face toward the side, and the teeth are big, with three cusps in a triangle on the upper teeth (figure 34). Above the orbits is a thick flange of bone, the supraorbital process. Those are features reported for a family of Eocene whales called protocetids, and this is the first protocetid of Kutch for which we have found a skull with partial skeleton. We call it *Dhedacetus*.[3] Protocetids are rarer than remingtonocetids, and given their size and the sacrum we just found, might easily have been the predators of the smaller, fish-eating, remingtonocetids.

PROTOCETID WHALES

Protocetid whales (figure 53) are found all over the world (figure 10) in rocks between forty-nine and thirty-seven million years old.[4] Their predecessors (pakicetids, ambulocetids, and remingtonocetids) only occur in India and Pakistan. It is thus likely that protocetids were the first whales able to migrate long distances and colonize the world. Protocetids are also the ancestors of basilosaurids, and through that, all later cetaceans, including the modern ones (figure 54).

Protocetids from India and Pakistan include a dazzling variety of genera: *Indocetus, Rodhocetus, Babiacetus, Gaviacetus, Artiocetus, Makaracetus, Qaisracetus, Takracetus, Maiacetus, Kharodacetus,* and *Dhedacetus*.[5] In addition, *Protocetus, Pappocetus, Eocetus,* and *Aegyptocetus* are known from Africa,[6] and *Georgiacetus, Nachitochia, Carolinacetus,* and *Crenatocetus* from North America.[7] Tantalizing fragments of a protocetid have also been discovered in Peru.[8] With so many genera, protocetids are more diverse and more widely distributed than any of the other families of Eocene whales.

More or less complete skeletons are known for only a few protocetids: *Rodhocetus, Maiacetus, Artiocetus,* and *Georgiacetus* (figure 55). Interestingly, forelimbs and hind limbs are missing in even some of these relatively complete skeletons, such as *Georgiacetus* and our *Dhedacetus*. It could be due to scavengers eating the cadaver, or because the cadaver was floating in the ocean, with pieces falling off and sinking as time progressed. With so many species for which no skeletons are known, it is difficult to know how representative what we know is for the entire group, but for now it seems that protocetids lived in the oceans like modern sea lions: hunting fast-swimming prey and powering their bodies with their limbs. However, they also had ties to the land, and probably went there for functions related to reproduction such as mating, giving birth, and nursing.

Feeding and Diet. Protocetids had powerful jaws and teeth, similar to ambulocetids, and different from remingtonocetids. The teeth and jaw

FIGURE 53. Life reconstruction of *Maiacetus*, a protocetid whale that lived in what is now Pakistan around forty-seven million years ago. Protocetids were the first whales to colonize the world's oceans.

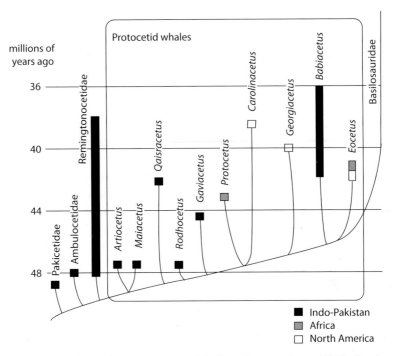

FIGURE 54. Relationships among some of the better-known protocetids. The basal and older protocetids are Indo-Pakistani, but later ones are known from most of the world's oceans. Phylogeny from Uhen et al. (2011).

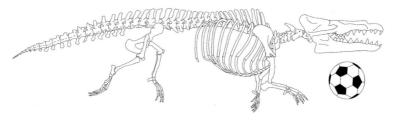

FIGURE 55. Skeleton of the protocetid *Maiacetus*, modified after Gingerich et al. (2009).

shape suggest that protocetids fought large and struggling prey. Isotopes of the teeth indicate that their prey lived in the water.[9] The protocetids from the Eocene of Kutch are much larger than the remingtonocetid whales that lived around them; as stated before, it is possible the latter were protocetid prey.

At first glance, all protocetids have similar dentitions. The dental formula is always 3.1.4.3/3.1.4.3. Upper molars have two large cusps and sometimes one additional cusp on the side of the tongue. Lower molars have a high trigonid in the front and a low talonid in the back, each with a single cusp (figure 34). In spite of that general similarity, there is ample evidence for dietary specialization between different protocetids. Most genera are like *Kharodacetus*: they have slender and high teeth with sharp edges and the distinct phase I attritional facets that characterize all early whales.[10] *Babiacetus* is different. The molars are more blunt and are worn apically, which implies heavy tooth–food–tooth contact. In one of the Indian *Babiacetus* specimens, one tooth in each left and right jaw are broken and only stubs remain. The two broken teeth are across from each other, and it is possible that both broke during the same violent jaw-closing event. Abrasional wear surfaces on both of these teeth indicate that the teeth were used after they broke. That means that the whale survived the damage. Unlike most protocetids, the left and right jaws of *Babiacetus* are fused with each other as far back as the second premolar.[11] This gave the animal a very powerful jaw, also suggesting that it fought even stronger prey than other protocetids. The ubiquitous marine catfish from Kutch have very hard, bony heads and may have been a prey item for *Babiacetus*.

The shape of snout and palate among protocetids varies greatly,[12] and those differences probably reflect dietary specializations, but this has not been studied in detail. The strangest protocetid face is certainly that of *Makaracetus*,[13] which has jaws that are not straight, but bent down. It certainly fed very differently from other protocetids, but we do not know on what or how.

Smell and Taste

Like the earlier whales, protocetids had a sense of smell. The olfactory organ of mammals picks up chemical compounds that are airborne, not those that are waterborne. It does this by trapping airborne molecules in the mucus on the inside of the nasal cavity, where these molecules stimulate neurons (nerve cells) that are part of a large nerve called cranial nerve 1, the olfactory nerve. These neurons send their findings to the brain via another nerve, the olfactory tract. Among modern cetaceans, olfaction is well developed in mysticetes[14] but appears to be rudimentary or absent altogether in odontocetes.[15] It is not clear why modern cetaceans use their sense of smell. Some modern whales (in the family Balaenidae, the so-called right whales) may detect the airborne scent of krill, which smells like boiled cabbage, to find their food. They have been observed to swim upwind when the smell of krill is in the air. The skull of the early whales indicates that members of this group had a sense of smell: there are small bony perforations for the olfactory nerve and a long bony tube for the olfactory tract in protocetids,[16] and the same is also true in remingtonocetids (figure 35) and basilosaurids.[17]

Olfaction in mammals is very different from taste. Taste receptors in mammals are mostly located on the tongue and palate. They are designed to detect chemical compounds suspended in a solid or fluid medium, and these signals are passed on to the brain via cranial nerves 7, the facial nerve, and 9, the glossopharyngeal nerve. Unfortunately, those cranial nerves do not have taste tracts that run in bony canals, and can thus not be studied in fossils.

There is a third chemical sense in mammals. The vomeronasal organ, also called Jacobson's organ, is located on the floor of the nasal cavity. It consists of a small sac that can detect large flavorant molecules through a duct that opens on the front of the palate. Not all mammals have a vomeronasal organ; humans lack it, for instance. In animals that have it, dogs for instance, the slit-like openings of its ducts are visible just behind the upper incisors on the palate. The vomeronasal organ plays an important role in detecting pheromones,[18] chemicals that are signals for other animals of the same species, such as those involved in sexual communication. In some artiodactyls, the vomeronasal organ is used to detect the reproductive status of conspecifics. Male deer engage in a behavior called flehmen, where they raise their head with open mouth, exposing the vomeronasal ducts on the palate to the air that, hopefully for the deer, carries molecules indicative of nearby females in estrus.[19] Like olfaction, the vomeronasal organ signals the brain via cranial nerve 1, and is stimulated by airborne molecules; but unlike olfaction, these molecules reach the organ via the oral cavity and are transported through the two slits on the palate. To get from the mouth to the nose, the vomeronasal ducts pass through two holes in the palate called the anterior palatine foramina, or, sometimes, the incisive foramen. No adult modern cetacean has a vomeronasal organ, and they also lack anterior palatine foramina. However, anterior palatine foramina do occur in pakicetids. Given that cetaceans are related to artiodactyls, it is likely that the vomeronasal organ was present in early whales. It was certainly lost when whales became more aquatic—in remingtonocetids and protocetids such as *Kharodacetus*—as indicated by the absence of anterior palatine foramina in their skulls.

We can only speculate what Eocene whales used their sense of smell for, but a reproductive function is possible. Sea lions have a sense of smell, and there too, it is important in functions related to reproduction: identification of potential mates, and recognition of their own young in a crowded colony.[20]

Vision and Hearing. A telling difference between protocetids and basilosaurids on the one hand and pakicetids, ambulocetids, and remingtonocetids on the other is the position of the eyes. Protocetids and basilosaurids have large eyes that face toward the side of the animal and are located below a thick, flat piece of the skull, the supraorbital shield (figure 52). These whales had excellent vision (unlike remingtonocetids) and used their eyes to look sideways, not just upwards like ambulocetids and pakicetids.

Protocetid ears are similar to those of remingtonocetids (figure 43): they show aquatic adaptations such as the enlarged mandibular foramen and the partial release of the petrosal from the skull, but these structures are not as perfected as they are in modern whales. It is likely that protocetids used their eyes and ears to hunt large prey and that most of that hunting took place below the water-line.

Brain. The deserts of Egypt have yielded a trove of casts of the inside of the braincase (endocasts) for basilosaurids,[21] and CT scans have revealed the shape of that cavity in remingtonocetids (figure 35). In Kutch, we found endocasts for the protocetid *Indocetus*.[22] Like all early whales, it has long olfactory tracts. There are also the beginnings of a rete mirabile, the network of veins that surrounds the brain (discussed in chapter 2). The retia (plural of rete) are largest on the left and right side of the brain, and there are none covering the top (dorsal) surface. In modern bowhead whales, the location of retia is similar.

In these Eocene endocasts, the front part of the brain (cerebrum in figure 35) is distinct from the back part (cerebellum). The relative dimensions of cerebrum and cerebellum are similar to those of land mammals and *Remingtonocetus*: the cerebrum is larger and higher than the cerebellum. It is different in basilosaurids. There, the cerebellum is much larger and higher, towering over the cerebrum. The surface of the cerebellar casts suggest that this space in basilosaurids is mostly covered with retia, greatly enlarged over their condition in protocetids, but it is still likely that the cerebellum makes up a greater portion of the volume of the brain than in other Eocene whales. In modern mammals, the cerebellum is involved in fine motor coordination, but we do not know whether there was a significant change in motor coordination between protocetids and basilosaurids.

The surface of the brain of all Eocene whales is relatively smooth, a condition called lissencephaly. In general, mammal brains that are larger (have higher EQs, see chapter 2) also have a brain surface with more and deeper grooves (sulci), and that relates in some broad way to how much brainpower they have. Eocene whale brains are different from modern odontocetes and mysticetes in that regard. That pattern—smooth brains in Eocene forms and convoluted brains in their modern relatives—is actually found in many mammal groups. In evolution, brain size and degree of convolution increased over the past fifty million years independently in different mammal groups.[23]

Unfortunately, it is at present impossible to determine what these early whales did with those brains. Brain organization in modern whales is very different from that in other mammals, and it is possible that this is related to increased cognitive skills and behavioral complexity.[24] However, we cannot know whether these organizational patterns occurred in Eocene whales. Their brains were certainly a lot smaller than in the modern forms.

Walking and Swimming. Protocetids such as *Maiacetus* had a robust vertebral column and short limbs.[25] Nearly all swimming mammals have short limbs. Sea lions, which propel themselves with their hands (figure 20), have short forelimbs and large, wing-like hands that can be forced through the water with powerful shoulder muscles.[26] Seals swim by means of pelvic oscillation. They have giant feet planted on very short thighs, and shins that allow powerful strokes in swimming. *Maiacetus* has short limbs, but its hands and feet are not large. Principal component analysis, a powerful mathematical method to study similarity in shape, has been used to study limb and trunk proportions of swimmers.[27] *Maiacetus* turned out to be most similar in skeletal proportions to the giant freshwater otter *Pteronura*, which is a caudal undulator (figure 20).

Indeed, the tail of protocetids is very interesting. The proportions of the vertebrae near the root of the fluke change abruptly in mammals with flukes (whales, dugongs) but not in those without flukes (otters, manatees; figure 12). Indeed, in *Maiacetus,* most known tail vertebrae are wider than they are high, but the thirteenth tail vertebra is higher than it is wide. In *Dorudon,* the thirteenth caudal vertebra is the ball vertebra and is located where the fluke hinged, and the height-width proportions of the vertebrae change here.[28] That suggests that *Maiacetus* had a fluke, and might mean that swimming using the modern means of fluke-driven caudal oscillation originated in the family Protocetidae.

Whereas the limb bones of *Pakicetus* and *Ambulocetus* are pachyostotic, this is not the case in protocetids. In *Pakicetus* and *Ambulocetus,* the extra bone is probably used as ballast to keep the animal underwater, which makes sense for hunting from ambush. In protocetids and *Basilo-*

saurus, the ribs are somewhat pachyostotic,[29] and these heavy ribs may have functioned as a stabilizer, as explained in chapter 2.

The limbs of protocetids had fully mobile joints, with well-developed fingers and toes, tipped by short hooves. Protocetids were certainly able to get around on land, although they were not fast or strong. The vertebral column of protocetids presents a puzzle. Nearly all mammals have seven neck vertebrae, and the number of thoracic and lumbar vertebrae adds up to the same number. Thus, the vast majority of mammals have twenty-six vertebrae in front of the sacrum (cervical + thoracic + lumbar vertebrae, together called the presacral vertebrae),[30] as pointed out in chapter 4. In fact, relatively stable numbers of presacral vertebrae occur in birds and reptiles too. The number of presacral vertebrae is also twenty-six in Eocene artiodactyls[31] and in most protocetids. However, in both *Ambulocetus* and *Kutchicetus,* the numbers are higher (thirty-one and thirty, respectively), and things get really out of hand in basilosaurids (*Basilosaurus,* forty-two; *Dorudon,* forty-one). Excess presacral vertebrae indicate that whales made a fundamental change in mammalian design, and the question is whether the number was increased twice in early whale evolution (in ambulocetids/remingtonocetids as well as in basilosaurids), or whether it was increased just once, with some protocetids reversing to the ancestral numbers.

Habitat and Life History. Pakicetids and ambulocetids were closely tied to freshwater, and remingtonocetids are common in muddy backbays. Protocetid fossils are often found in deposits indicative of clear, warm, and bright waters (figure 30).[32] Such seas sustain ecosystems with diverse life-forms, and protocetids were probably the top predator of these systems. Although most protocetid fossils have been found in such near-shore but open marine environments, it is likely that they also inhabited the surface waters of the deeper oceans. Those environments do not easily fossilize, and less is known about diversity there. It is possible that the oceans teemed with cetacean life soon after protocetids appeared on the scene.

Still, it is also clear that protocetids retained ties to the land. If seals and sea lions are a modern analogue for protocetids, it could be that functions related to reproduction required a stable substrate. Of course, those functions—mating, birthing, and nursing—do not fossilize easily. In general, fetuses and newborns have bones that are soft and fossilize poorly. A small whale inside a larger whale's body was discovered for the Eocene whale *Maiacetus*—a remarkably beautiful specimen. The head of the small one is facing toward the tail of the larger individual, and it has been interpreted as a fetus inside the mother's body.[33] However, the smaller individual is located where the mother's heart and stomach used to be, not where her uterus was. Modern baleen whale fetuses are commonly found in the chest cavity after the death of the mother, when rotting gases in the abdomen propel the dead fetus forward, through the diaphragm and into the chest. Furthermore, the skeleton of the little whale is incomplete; its entire back half is missing. Is it possible that the adult killed a small free-swimming specimen, biting it into two pieces and swallowing one part. The bones of the small specimen are so undefined that one cannot determine whether these two are even the same species. There are ways to study this further: bodies of fetuses are physiologically part of their mother's body, so if the isotopic signature of the small specimen matches that of the adult one, the mother–fetus relation seems more plausible; but that work has not been carried out yet.

PROTOCETIDS AND HISTORY

The first protocetid was discovered in 1904 in the desert of Egypt.[34] The site where it was found is now gone: the city of Cairo has expanded over it. That specimen was a skull. It was named *Protocetus atavus,* Latin for "before-whale grandfather." It was immediately recognized as a possible link to land mammals, and for nearly a century it defined what people thought an ancestral whale would look like. But it was just a skull, and scientists did not realize how different protocetids really are from modern whales. The skull survived more than forty million years of burial in Africa, but not four decades of being housed in a natural-history museum in Stuttgart, Germany: it was destroyed during bombing in World War II.

Protocetids are fascinating whales—the first ones to disperse across the planet, adopt the fast-hunting strategies that many modern whales still use, and reach unprecedented levels of diversity, both in numbers of species and in morphology. Having said that, they cannot help me understand what the ancestors of whales, the critters before pakicetids, looked like, and they cannot solve the riddle of the relation to hippos. For that, I need to study artiodactyls—old ones, and preferably from India or Pakistan, since that is where cetaceans originated. Again, I am confronted with the fact that I have to focus away from marine rocks and start digging in rocks that have terrestrial animals.

CHAPTER 13

From Embryos to Evolution

A DOLPHIN WITH LEGS

Tokyo, Japan, June 7, 2008. No living cetacean has legs that stick out of its body. Except for one, and I am in Japan to see it: a dolphin with hind limbs. I have seen pictures of the animal on the Internet, showing two triangular fins emerging from the body near the slit where the genitals lie hidden. The animal made headlines around the world, and my Japanese colleagues offered to take me to see it.

The dolphin's capture is controversial. Tadasu Yamada, who studies whales at Japan's National Museum of Nature, tells me that the dolphin was caught by dolphin hunters from the village of Taiji, about 300 km west of Tokyo.[1] These hunters are infamous for their practice of scaring groups of dolphins into narrow coves with loud sounds, and then killing them, apparently for food. This one dolphin looked different, and the hunters kept it alive, housing it in a nearby marine park. They call it Haruka, which means "coming from ancient times," a reference to the evolutionary origin of hind limbs.

I visit Tadasu in his office in Tokyo with Jim Mead, the curator of marine mammals at the Smithsonian Institution in Washington, D.C. Both of these senior anatomists delight in anatomical trivia, launching with gusto over lunch into a detailed discussion of the anal tonsils of cetaceans—where they are, what they're for.

After they exhaust that topic, Jim says to me: "Back in the sixties, I worked east of Tokyo, in Chiba Prefecture. There was a whaling station in Chiba, and we would stay there and study the beaked whales they pulled up, *Berardius*."

Berardius is the Latin name for a species of beaked whale that lives in very deep water. This whale is enormous and eerie, with big eyes—something like a monstrous Flipper crossed with a giant squid.

We talk about our past experiences studying whales in Japan. Japan is a big whaling country, but whaling is regulated by a group called the International Whaling Commission (IWC). Japan exploits a loophole in the IWC regulations, allowing scientific whaling, and kills thousands of whales in the name of research. But few scientists—including Japanese scientists—are impressed with this research.

Any nation can join the IWC and have a vote. The meetings are often aggressive, with whaling nations such as Japan, Norway, and Iceland going head to head with conservation-minded nations like Australia and New Zealand. The four-legged dolphin is an example of scientific research on the coattails of hunting, but it's a very unusual one. We talk about how to get access to it. Yamada had emailed me that the Japanese scientists at the Taiji whale museum wanted Jim and me to give short presentations, explaining what they could do scientifically with the animal. I was excited, and had called Frank Fish about the possibility of experiments. Frank got excited, too. Two days later, I received word from Tadasu that the whole thing had fallen through. The administrators running the aquarium wanted nothing to do with outsiders. There would be no talks, and we would only be allowed to see the animal in the tank, as if we were tourists, with no special access. I ask him why.

"They want to keep the animal alive as long as possible, and breed it. They have put her in a tank and want nothing done with her."

I hadn't planned any invasive or damaging experiments, mainly just filming the animal. But I can see that this is not simple.

"Have they CT-scanned the animal?" I ask.

"No, they really limit all handling. They have made clear that they only will allow access to the specimen to people who openly support the drive hunt."

That counts me out, because I do not. "How can we help other scientists get access? I presume that they only want Japanese to study the animal."

"No, foreigners are fine, as long as they support the drive hunt." Studying Haruka has become a political act.

The three of us meet three more Japanese, and together we fly to Wakayama Prefecture, where the village of Taiji is located. Ours is the only plane. The runway is hewn out of black rock that erodes in steep cliffs, and it is so narrow that the plane taxis to the terminal on the runway—there's no room for a taxi lane. A big whale is carved into the hillside across from the terminal building, and an airport store sells canned whale meat. On the two-hour drive along the rocky, winding coast, we stop at a restaurant which sells whale-shaped incense burners and piggy banks. Later, we drive by a giant waterfall in the shape of two whale flukes.

It finally dawns on me that this is the heart of Japanese whaling country. Most Japanese don't eat whale meat, and I've never seen it on any of my many previous trips to Japan. However, the government tries to encourage its consumption, and in some areas, such as here, whale and dolphin hunting is part of the local culture. These people fiercely protect it.

THE MARINE PARK AT TAIJI

The next morning we go to the Ocean Park, where the skeleton of a large blue whale is mounted outside. The white bones contrast with the black rocks that frame the valley, creating a very Japanese image, ready for a woodcut artist to immortalize. A curator meets us, but all the conversation is in Japanese, and Jim and I just walk along. Then the director—the man who apparently pulled the plug on our talks—comes out. He too is very courteous, and introduces us to the veterinarian and trainers. Business cards are exchanged in a very Japanese ritual, and the stack of my own cards diminishes as my stack of received Japanese cards rises higher and higher. The director's card has a picture of Haruka, and I realize that this is an important attraction for them. We walk past a large building with two whales painted on it and along a sea arm that reaches between the cliffs. As I look at the water, a large black fin emerges from it, followed by a loud *whoosh*. It startles me, but I recognize that a killer whale has come up to breathe. Why is such a large animal swimming in this narrow cove? I scan the bay and realize that the entrance to the cove is dammed, and forms a natural aquarium for their captive killer whale. Farther down, there are rectangular wooden structures floating on barrels in the water. The water inside occasionally ripples with something dark: these are pens with cetaceans in them. I count at least six pens, some with multiple animals. We walk down a winding path to a round concrete building, built around a large tank. As we walk in the door, I realize that we are in a giant plexiglass tube at the

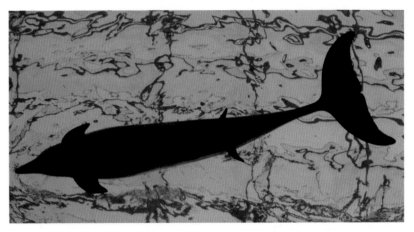

FIGURE 56. Haruka, the dolphin with small rear flippers, seen from below. The rear flippers are its hind limbs. Normal dolphins do not have rear flippers; the structures developed in this individual as a congenital anomaly.

bottom of the aquarium, and three bottlenose dolphins are lying on top of the tube. One is Haruka (figure 56). We all crowd around with our cameras. The dolphins are not bothered. Two of them start to play. The third stays; it is Haruka. She seems to know that she is the star of the show, and wants to please the fans with photo ops.

Haruka has real little flippers in the back, one slightly larger than the other, but nicely formed, like miniature front flippers. The trainer says that one of them is much more loosely attached than the other. The director, the veterinarian, and the curators have stayed with us. They motion us away: "Please, please." Do they want us to leave because we have antagonized them, I wonder?

No, quite the opposite. They lead us to the roof of the building, to the top of the aquarium. The trainers take coolers and go onto a platform in the tank, and the dolphins show up immediately. The trainers feed fish to the dolphins. One of the trainers gestures, and Haruka rolls over, exposing her belly. The trainer gently puts his hand underneath the flipper and we shoot pictures and video. All captive dolphins are taught to go belly-up so that a trainer or vet can examine their genitals and anal region, and take their temperature rectally. It lasts a few minutes, and then the dolphin receives another signal, rolls over, and loudly snorts. When she is on her belly, she cannot breathe. She gets another fish, and is told to roll over again, to the sound of more camera clicks and more movies. An hour passes before I realize it.

SHEDDING LIMBS

Haruka is not the first modern cetacean born with hind limbs. There are more than half a dozen reports of whales and dolphins with such structures,[2] varying in size from the little bumps on the abdomen of a Russian sperm whale to the four-foot-long appendages in a humpback whale caught near Vancouver Island, Canada, in 1919.[3] But unlike all of the others, Haruka is alive.[4] She might be able to teach us how hind limbs affect swimming, and why whales lost their hind limbs in the first place.

To understand why Haruka is different, we need to understand how limbs develop in other mammals. Limb embryology is relatively similar across vertebrates, from fish to birds to mammals. Early in development, an embryo looks a lot like a worm, with a head, segments along its back, and a tail, but no limbs. Human embryos look like this too, until in the fourth week after fertilization, when the embryo is still smaller than a pea, it grows two small bumps in the chest region. Later in the fourth week, two small bumps appear at the base of the tail. Limb buds form in this way in all vertebrates, but there are many differences too. Timing is one: in a mouse, limb buds form much earlier, around day 10. To make it easier to compare embryos of species with different gestation times, embryologists have divided development into so-called Carnegie stages. Limb buds in most mammals start forming at Carnegie stage 13.[5]

Limb buds consist of two kinds of cells. The outside is covered by a single layer of flat cells that cover the embryo like the pavement on a street. These are called epithelial cells. Inside, the entire bud is filled with undifferentiated cells that make up the mesenchyme. The epithelial cells on the top of the limb bud form a crest called the apical ectodermal ridge, or AER. The limb bud grows, and inside of it clusters of cells clump together (or condense) and form cartilage bars that will turn into bone. One such bar forms between shoulder and elbow—it will be the humerus. Two bars form between elbow and wrist: the radius and ulna, the lower arm bones. Five distinct bars of cartilage take shape in the hand, the precursors of the fingers. In most mammals, roughly the same bars form in the lower limb, to make the leg and toe bones. Once the cartilage bars have formed, cells that will form muscles migrate into the developing limb from the body. Initially, there are no separate fingers in the hand or distinct toes in the foot. The cartilage bars are embedded in a flat pad of tissue, and the hands and feet look like mittens without even a thumb. As the hand and foot develop, the tissue between the

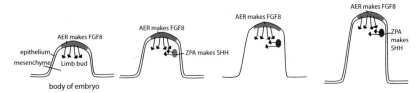

FIGURE 57. Diagram of embryology of the limb in a vertebrate. The forelimb and hind limb initially form as a small limb bud (left diagram) that projects from the body wall. Over time (drawings further right), the limb bud will grow, and eventually a skeleton will form inside it. AER, apical ectodermal ridge; FGF8, fibroblast growth factor 8 (a protein); ZPA, zone of polarizing activity; SHH, sonic hedgehog (another protein).

digits thins and eventually disappears to make five independently movable fingers and toes.

The genes regulating all this are reasonably well known (figure 57). Initially, the AER produces a protein called FGF8,[6] which leaks into the mesenchyme underneath it. An area like this, which produces a protein that is used to communicate to other areas, is called a signaling center. A group of mesenchymal cells in the rear of the limb bud also becomes a signaling center, the zone of polarizing activity (ZPA). The ZPA now starts to produce a protein called sonic hedgehog (after the videogame character), a name usually shortened as SHH. SHH diffuses into the tissue around it and is necessary to keep the AER alive and working at this stage. The mesenchymal cells immediately underneath the AER divide and as the limb bud grows longer and longer, cartilage bars form and these divide the limb into segments: in the forelimb, shoulder to elbow, elbow to wrist, and hand; in the hind limb, hip to knee, knee to ankle, and foot.[7]

The ZPA plays a role early on, working with the AER to accomplish limb-bud growth. The ZPA again plays a role later in development, during the formation of the fingers and toes. SHH produced by the ZPA oozes into the surrounding mesenchyme, and, because the ZPA is located on the pinky side (posterior side) of the hand (the little-toe side of the foot), concentrations of SHH drop as you go toward the thumb (big-toe) side. As growth continues, cells farther toward the thumb side will receive both a lower concentration and a shorter duration of SHH exposure. That is the signal that makes different fingers and toes: the index finger is exposed only briefly to low concentrations of SHH, the pinky for a longer time and higher concentrations, and the remaining fingers are intermediate.[8] This controls the particular shape of the fingers and toes.

In many mammals, forelimbs and hind limbs follow the same trajectory, but this is not true in cetaceans. In the forelimb,[9] development initially proceeds as in most other mammals, until the soft tissue that connects the digits fails to disappear and make separate fingers. In addition, unlike many other mammals, many cetacean species form more than three phalanges per finger or toe (figure 13). This all makes a smooth, asymmetrical flipper, a blade that cuts through the water and is used in steering.

The trajectory for the cetacean hind limb is quite different from that of the forelimb. Although, in living cetaceans, external hind limbs appear only if development goes wrong—in animals like Haruka—every living (and presumably fossil) cetacean had hind limb buds as an embryo (figure 58). Those buds form but eventually disappear long before birth, as their developmental trajectory is cut short. This leaves only some internal structures that are attached to the genitals (figure 15).

Hind limb buds in cetacean embryos were discovered a long time ago, and that story holds some lessons for us in the present. In *The Origin of Species*, Darwin did not make much of embryology's contribution to evolution, but mainland European embryologists enthusiastically adopted his ideas about evolution, because these allowed them to interpret previously inexplicable features of embryos. In 1893, three decades after Darwin's book, the German embryologist Willy Kükenthal interpreted the two bumps low on the abdomen in a porpoise embryo as hind limb buds. Kükenthal's publication[10] was much to the chagrin of his Norwegian colleague Gustav Guldberg, who had talked about hind limb buds in prenatal cetaceans in a lecture a few years earlier. However, Guldberg's buds were not in the same place as those of Kükenthal, and also at a different time in development. Guldberg, who worked in a small institute in Bergen, off the academic beaten path, had failed to publish his finding. He felt scooped by the professor from the famous university in Jena, but remained civil: "I was thus rather surprised . . . that in the work . . . of my friend Professor Kükenthal, . . . he appears to believe to show the presence of the early beginning for hind limbs in a porpoise embryo of 25 mm length." Kükenthal's bumps were too far forward on the body to be hind limb buds, according to Guldberg, and instead he described buds in additional porpoise embryos of 7, 17, and 18 mm that document earlier stages of development.[11] Kükenthal dismissed Guldberg's buds as the beginnings of the mammary glands instead of the limbs, writing: "Grave reservations arose in me, when I read that Guldberg considers the hind limb buds to be two prominences in embryos of 17 and 18 mm."[12]

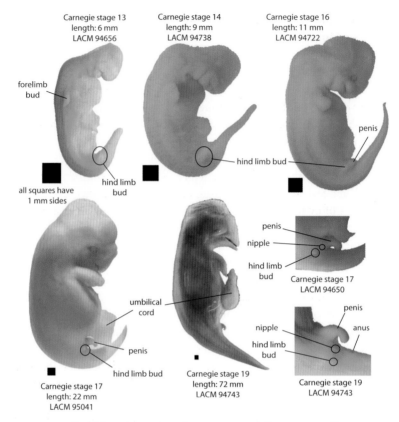

FIGURE 58. Dolphin embryos (*Stenella attenuata*) at different stages of development and not at the same scale (black squares have 1 mm sides for all specimens). These photos show the increase in size of the forelimb bud and its development into a flipper. They also show the increase in size of the hind limb bud (first and second photo) and its subsequent reduction in size, and disappearance. Nipples in dolphins are located next to the genitals, and they are visible in the lower two photos (and their enlargements on the right). Embryonic stages are referred to as Carnegie stages. Length of the embryo (CRL, crown-rump length) is a good indicator of age as embryos grow rapidly.

The confusion is not as silly as it seems. In an early embryo, a budding hind limb and a mammary gland are not so very different. Both begin as a swelling filled with mesenchyme and covered by epithelium, with one patch of the epithelium thickened. In addition, the nipples of modern cetaceans are in fact located on either side of the genitals, near where limbs would be. Land mammals such as moles and squirrels have nipples in their groin, and these form in the embryo as low welts between genitals and hind limb.[13] However, nipples start to form long after the limb buds. Human nipples form in the seventh week of development, during Carnegie stage 17.

Guldberg fought back, now with an exhaustive description.[14] He had hoped that more embryos would become available, but since they did not, he was left restudying those that he had already. "As the short essay by Prof. Kükenthal may have cast a dark shadow over my work ... detailed restudy confirmed *my* findings in all directions" (Guldberg's italics). That is where the conversation ceased, until Marga Anderssen, also Norwegian, confirmed Guldberg's interpretation some twenty years later.[15] She studied a larger collection of embryos, covering a greater part of porpoise development. She confirmed beyond doubt that the outer welt, the hind limb bud, develops earlier and disappears eventually, whereas the inner welt begins its development later, and eventually becomes the mammary gland.

The broader lesson is that these ancient embryologists were hampered by the lack of material. They had only a few embryos. What they needed was an ontogenetic series: embryos that cover all developmental stages of one species, from some that are just a few millimeters in length to those that already look like a miniature of the adult (in cetaceans, these are the size of a mouse). Critical embryos, those between 7 and 17 mm, were missing in the collections of both Guldberg and Kükenthal.

Knowing about the hind-limb loss in evolution, and reading about their embryology and the genes controlling it, has me psyched. There is some real potential here to understand evolution at a deeper level. Apparently, the genetic program that makes hind limbs in other mammals is still operational in cetaceans—starting up as usual in early development, but then grinding to a halt a few weeks later. Add to that the great variation among modern cetaceans. Dolphins have just one bone near the area of the hind limb, the pelvis, whereas others, like bowhead whales (figure 59), have pelvis, femur, and tibia, all embedded internally. That suggests that the developmental program is switched off at different times in different whales. With so much information on the genetics of limb development, and spectacular fossils documenting the transition, it would be amazing if one could figure out which of the limb genes were altered in development, and when that evolutionary change happened in cetacean history. But how do you get an ontogenetic series of cetacean embryos?

It is Bill Perrin, a scientist at the National Oceanic and Atmospheric Administration in La Jolla, California, who points me in the right

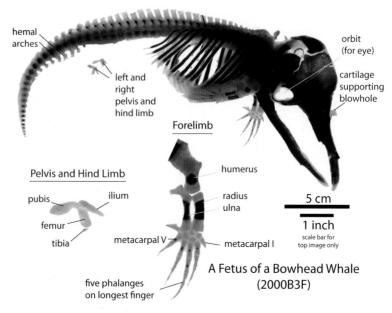

hemal arches

left and right pelvis and hind limb

orbit (for eye)

cartilage supporting blowhole

Forelimb

Pelvis and Hind Limb

pubis

ilium

femur

tibia

metacarpal V

humerus

radius
ulna

metacarpal I

five phalanges on longest finger

5 cm

1 inch

scale bar for top image only

A Fetus of a Bowhead Whale (2000B3F)

FIGURE 59. Bowhead whale fetus, treated so the soft tissues are transparent, bone is purple, and cartilage is green. This technique is called clear-and-stain. Compare the hand bones to the hand of an adult whale, in figure 13. This fetus also shows the pelvis and hind limb, which are embedded in the abdomen in life (see figures 14 and 15). Baleen will form in the gap between upper and lower jaw.

direction. Bill was the person who, back in the 1980s, started to sound the alarm bell over how many dolphins died in the tuna fishery. He initiated a movement that eventually made the world a safer place for dolphins. In the process, scientists got access to many of the dolphins that drowned in tuna nets, and embryos of the pregnant ones were collected. The embryos ended up with whale scientist John Heyning at the Natural History Museum of Los Angeles County. When I contact John, he is immediately intrigued by my question about the genes that form the dolphin hind limb, and off we go to the museum's warehouse, where entire ontogenetic series of several dolphin species are housed in little vials of alcohol. It is just the resource we need.

To study which proteins are active at what time in development, we cut embryos of all stages into very thin slices and search them for specific proteins involved in making limbs. A pea-sized embryo yields about a thousand slices, each 7 microns thick. Such slices are mounted on glass slides and are so thin that light shines through them, so they can be stud-

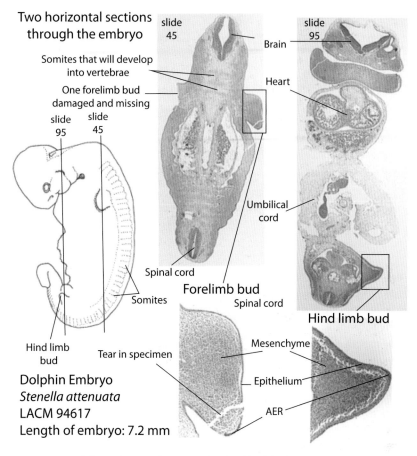

Two horizontal sections through the embryo

slide 45

slide 95

Brain

Somites that will develop into vertebrae

Heart

One forelimb bud damaged and missing

slide 95   slide 45

Umbilical cord

Spinal cord

Forelimb bud

Somites

Spinal cord

Hind limb bud

Hind limb bud

Tear in specimen

Mesenchyme

Dolphin Embryo
Stenella attenuata
LACM 94617
Length of embryo: 7.2 mm

Epithelium

AER

FIGURE 60. Dolphin embryo at the stage when hind limb buds are largest. Lines mark areas where thin slices (shown on right) were taken from this specimen. These slices are affixed to microscope slides. Two of such slices are shown with enlarged areas showing the sectioned hind limb bud. As development proceeds, the hind limb buds will regress and disappear.

ied with a microscope (figure 60). The organs can be easily recognized now, and the limb bud and its AER studied. Proteins made by the embryo, FGF8 for instance, are embedded in the organ in which they were initially, now stuck to the glass slide. Then we douse the slice with another protein, an antibody, designed specifically to search out and bind to the FGF8 protein and only that protein. A special dye binds to the antibody and reveals, in brown, areas where this FGF8-antibody complex is located. Other dyes (histologists call them stains) are used to color the

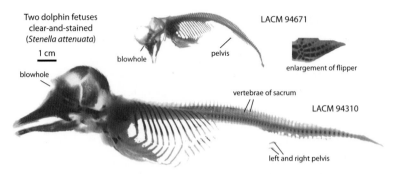

FIGURE 61. Clear-and-stained dolphin (*Stenella attenuata*) fetuses. Note that the smaller individual has less bone (purple). The photo of the flipper shows that all bones of the hand are already present even though this fetus is tiny and long before birth. See figure 59 for explanation of the technique.

tissues in purples and pinks, blues and reds, until the tiny embryo slice looks like an expressionist painting, gorgeous and full of information.

Working with our dolphin embryos, we find FGF8 in the AER of both forelimb and hind-limb bud. That was expected, of course. We next use an antibody for SHH, and it indicates the presence of a ZPA in the forelimb, but no SHH-producing ZPA in the hind limb. Thus, in the hind limb, SHH is a broken link in the chain of genes needed to make a limb. We conclude that the AER was present and functional initially in both forelimb and hind limb, but that it dies prematurely in the hind limb of dolphins as a result of the absence of an SHH-producing ZPA.[16]

That observation in dolphin embryos helps to explain why different living cetaceans vary in their limb bones. In dolphins, the pelvis remains, but no hind limb bones are left (figure 61). That indicates that the AER dies early, the result of the total absence of SHH in the hind limb bud. Bowhead whales, however, always have a femur, a piece of cartilage or bone that represents a tibia, and occasionally even parts of a foot (figure 15). That could be explained by the longer life of the AER, which may keep going in the presence of the ZPA for a longer time. Of course, you'd need some bowhead whale embryos to prove the idea.

If SHH is a protein that changes over the course of whale evolution, we might also use it to help us understand the hind limb shapes of fossil whales. *Basilosaurus* from chapter 2 comes to mind, with its tiny hind limbs that still contained femur, tibia, and fibula, plus three toes with two phalanges each. So how can SHH contribute to that foot shape? As we saw, SHH helps direct mesenchyme to form into fingers and toes.

The three-toed feet of *Basilosaurus* remind me of certain lizards called skinks, which were studied by a developmental biologist named Mike Shapiro.[17] In some species of skinks, the number of digits varies between individuals: there can be two, three, or four. Mike found out that more fingers and toes form as the hand and foot of the embryo are exposed longer or to higher concentrations of SHH. So we know that SHH plays a role in the unique hind limbs of dolphins, and we know that its absence causes loss of toes experimentally in mice, and in nature in skinks. Taken together, all this suggests that the reduction of toes in *Basilosaurus* more than forty million years ago was caused by a drop in SHH in the foot, a decrease that foreshadowed the loss of all of the hind limb elements in younger whales. The data from genes, embryos, and fossils complement each other beautifully here and are able to explain the evolutionary pattern and the modern shapes of the cetacean hind limbs.

Interestingly, during the part of Eocene whale evolution before *Basilosaurus,* there were great variations in the shape of different parts of limbs among whales. That variation was related to function in chapter 4. However, in spite of that variation, Eocene whales always retain the femur, tibia/fibula, and foot, even as the function of hind limbs in locomotion is lost. So the function of the hind limbs was reduced with no changes in SHH action early on. This indicates that the very fundamental changes in the limb development process that eventually reduced the hind limbs to nothing in the modern species were actually not driving the significant functional changes related to different locomotor behaviors (figure 62).

Some developmental biologists believe that evolution is driven by minor genetic changes very early in the embryo. That is based on the understanding that such changes have the opportunity to modify the resulting individual in very fundamental ways. However, in the case of whale hind-limb evolution, the ontogenetically early developmental change did not occur in evolutionary time until their locomotor function had long been lost.

Given all this knowledge from experiments and nature about SHH, one wonders what happened to Haruka, the Taiji dolphin. She appears to have the bones that are normally present in land mammals: femur, tibia, fibula, and some ankle bones. There are two or three toes, one with multiple bones. The pattern is roughly similar to that of *Basilosaurus* and skinks, suggesting that some mutation involving when and where SHH was expressed, when Haruka was an embryo, helped to create its hind limbs.[18]

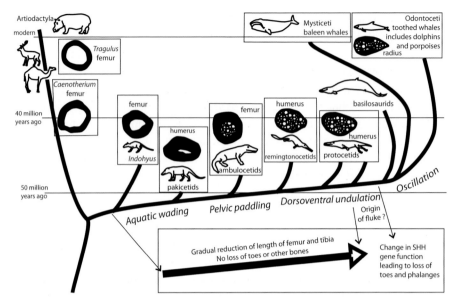

FIGURE 62. Cladogram summarizing evolution of swimming modes in early whales, with changes in bone density, and inferred change of SHH gene function. Irregular ovals are cross-sections of bones; black indicates dense cortical bone; lattice is spongy bone; and white is the marrow cavity.

Haruka also reminds us that the developmental process that makes hind limbs in other mammals is still locked into the cetacean genome; it is just switched off in normal individuals. Maybe Haruka's AER kept going, long enough to make small rear flippers. It would be amazing if we could sequence the DNA of this animal and compare it to that of other dolphins. It is unlikely that the nucleotide sequence of SHH differs between normal dolphins and Haruka, since that protein is involved in so many processes. A mutation in it would probably cause lethal deformations in the embryo. It is more likely that some gene involved in switching the SHH gene on and off in the hind limb is different, causing a difference in the timing of SHH expression. Those genes are called regulatory genes.

If we could sequence Haruka's entire genome, there would be thousands of differences with other bottlenose dolphins, and most of those would not be related to hind-limb development. However, the differences that did cause its hind limb bud to keep on growing would be there too. And on those regulatory genes would be the fingerprints of cetacean evolution. The same regulatory genes may also have effects on other parts of the dolphin's anatomy, and possibly those same genes were involved in shaping the other parts of the anatomy of the Eocene cetaceans so that different features that were evolving in cetaceans were not inherited independently. These thoughts go through my head as I think about Haruka. As DNA-sequencing technology gets cheaper and faster, it should not be that hard to sequence bottlenose dolphins and figure out what their differences are. And as we learn more about development, it would make a really compelling story when combined with the fossil evidence. Of course, one would need access to DNA from Haruka. And that, I do not have.

WHALING IN TAIJI

Back in Taiji, our visit with Haruka ends. In the main building, the director of the museum goes to the gift shop and brings each of us a tie clip in the shape of Haruka. A poster for the aquarium features her as its main attraction. Outlines of a series of fossil whales—*Pakicetus, Ambulocetus, Rodhocetus, Basilosaurus*—grace the bottom of the poster. I cannot read the Japanese text, but it is exciting to see that the whales that I know so well have made it to this very small town off the beaten path, and that in this place, everyone cares about whale evolution. The curator takes us out for lunch—there is whale meat—and then shows us the sights of Taiji. We see where dolphins are kept for shipment to aquariums, and the cove where the others are killed. I remember videos of fishermen killing the animals by stabbing them with long knives. Sad and infuriating.

I think back on a display on the history of Taiji in the museum of the marine park. Taiji was a little town on an unfriendly, rocky coast. There is no flat land to grow crops, and in the past, this village was connected to other villages only by slow winding paths. Coastal commerce was not possible, because too many submerged rocks defeated cargo ships. Going out in small boats to catch seafood was all the people could do. Whale and dolphin catches were initially opportunistic, but around 1600, whaling became an industry, with men stationed on high points as lookouts, using flags to signal wooden rowboats that were already out on the ocean waiting, so a whale chase could start immediately. Boats with about eight rowers gave chase, driving a large whale into a net and slowing it down. A hunter with a harpoon would stand on the stern of a row boat, throwing his weapon as they got close enough. This was a major undertaking and might easily involve twenty boats. On a hill nearby, we visit a monument to the whaling disaster of 1874. The town was starving, and a whale was sighted. The whale had a calf, and the whalers did not normally pursue mothers with calves. Emotions ran high among the hunters, with different factions pushing to get the whale so it could feed the people, or urging that it be left alone. A storm was approaching. The empty-belly argument won, and the tiny wooden rowboats went out. The storm closed in on the hunters. For days afterward, the bodies of young men, old men, and boys floated ashore: 111 did not return alive.

Back from my trip, in Tokyo, I present my research findings on our study of the development of hind limbs in dolphins. Still jet-lagged, I am sleepless at 3 A.M. as I go through the program. The talk after mine will be by Seiji Ohsumi, the director of the Cetacean Research Institute, a government agency that coordinates the whaling in Japan. His institute is the face of the Japanese whaling industry to the outside world. The industry claims that they catch whales for study—"scientific whaling," a rather laughable cover-up for the Japanese commercial whale-meat industry that fools no one. Ohsumi and his colleagues are considered the devil in a kimono by the whale conservation world. Ohsumi's talk is called "A Bottlenose Dolphin with Fin-Shaped Hind Appendages." I consider that he will probably see my presentation.

I would like to study Haruka, but I cannot agree with the drive hunt. On the other hand, my information that consent with the hunt is a condition is second-hand. If I did work on Haruka, who was captured in the hunt, would that imply approval of the hunt?

I add a few of my pictures of Haruka to my presentation, labeling them so it is clear where I took them. I also get ready to mention what may have caused this particular anomaly in this individual. I am speculating that SHH was switched off later in development than in other cetaceans.

Ohsumi speaks after me. He is an old man, with blotched skin and small eyes, well past seventy, I would guess. That generation of Japanese is formal, and he wears a gray jacket, whereas most of the conference attendees have shed their jackets after the first day. He discusses the school of dolphins that was caught on October 28, 2006. Some escaped as it was driven to the cove, but 118 dolphins were captured. Of these, ten were "kept," to be used for dolphin shows and aquaria, and the implication is that the rest did not leave the cove alive. Haruka was one of the lucky ten. Ohsumi goes through the other known cases of anomalous development of the hind limbs of cetaceans, including the humpback whale caught near on Canada's west coast in 1919 and the Russian sperm whale. Then he shows a diagram of the management and study of the animal, including a Breeding Group, a Function Group, a Genetics Group, and a Morphology Group. Finally, he encourages those interested in studying the animal to contact an e-mail address on the screen. I start to write it down: "haruka@"—and then stop. I cannot be part of this.

Later, I speak to a Japanese scientist who explains that the Institute for Cetacean Research is not on good terms with most Japanese academic scientists. The academic scientists do not believe that the scientific whaling research yields trustworthy data. They find that the outcomes are driven by political motives.

"In Japan, there are two sides, and the whaling gives Japanese science a bad name," he says.

I consider the fact that the entrails of all these dead dolphins, including their early embryos, still in the womb, are lying on the black rocks near where I was a few days ago. It's an incredible opportunity, and a thoroughly disturbing image.

I think about the nuances of the issue. Dolphins are different from whales—smarter, more social. At some point, the mayor of Taiji proposed to stop the dolphin hunt if his town was allowed to hunt fifty minke whales. Whaling of a species that is abundant, not all that intelligent, and killed after a fast chase, seems more humane and sustainable than the dolphin hunt. From my perspective, it is an idea worth considering. However, the IWC is like a dysfunctional family, and the pro- and contra-whaling groups are too far apart for a compromise. There are no winners in this fight. Everybody loses, including the whales.

Haruka will live her life,[4] receiving excellent care in her golden jail and providing propaganda for the Japanese whaling industry. She will hopefully also inspire its visitors to be engaged about whales. Maybe some good will come of that.

To learn more about hind limb evolution in cetaceans, I need to study artiodactyls—old ones, and preferably from India or Pakistan, since that is where cetaceans originated. Again, I am confronted with the fact that I have to focus away from marine rocks and start digging in rocks that have terrestrial animals.

CHAPTER 14

Before Whales

THE WIDOW'S FOSSILS

Driving on the Gangetic Plain in India March 12, 2005. It is a long and pleasant drive to Dehradun, a straight road initially, then suddenly the Himalayas appear at the horizon. An hour later, the road, a lane-and-a-half wide, reaches them, and snakes across their front range. Today, we are traveling in the middle of an artillery convoy, trying to pass the trucks one by one. The passing is useless. In front of each truck–cannon pair is another truck–cannon pair. It seems as if all the guns the Indian army has are being moved to Dehradun. We reach a tunnel through a mountain, and come to a stop, the barrel of the gun pulled by the truck in front of us pointing straight at our windshield. "I hope that it is not loaded," my assistant, Brooke, says. In quiet, I wonder if it is an omen that predicts fireworks to come on our mission.

I am on my way to meet Dr. Friedlinde Obergfell, the widow of the Indian geologist Anne Ranga Rao. Ranga Rao discovered a rich fossil locality near the Line of Control in the Himalayas, in the disputed territory of Kashmir, near the village of Kalakot. It turned out to be the largest collection of Eocene fossil mammals known from this subcontinent, larger than German professor Dehm's sites in the Kala Chitta Hills, and those of all fossil collectors that were here before: West, Gingerich, and myself. Ashok Sahni, the heavyweight of Indian paleontology, heard about it, and sent his student to the site to collect too. Ranga Rao was furious—his

locality was being raided. He was also rich, and he had the entire site excavated. Trucks were loaded with the fossil-containing rocks and took them
to his estate in Dehradun. Professor Dehm invited Ranga Rao to come to
Germany and study Eocene fossils. There, he met Dehm's assistant,
Friedlinde Obergfell, and married her. Ranga Rao was an outsider to paleontology, unable to study and publish his fossils adequately. The experience with Sahni drove him and his wife into seclusion and secrecy. Although
he was able to extract a few fossils and publish them, most were left in
burlap bags in his cellar, and a mound of fossil rocks occupies his yard.
The paranoia worsened; he drank and chain-smoked, and eventually died
from a brain tumor. After his death, his wife, left alone in a country where
she did not even speak the language, kept up the siege mentality. She
allowed no scientist to study the fossils, and approached even the most
innocent interactions with paleontologists with the greatest suspicion.

In spite of that, I have been trying my luck with her by visiting her
every time I come to India. I hope to endear myself to her, being European by birth and being able to speak German. I have my reasons. This
is the largest collection of Eocene artiodactyls from India, and it is our
best bet at finding the closest relative to whales. These artiodactyls are
in the right place, at the right time, a fact that has not escaped other
whale workers, too.[1]

With Brooke and several Indians, I visit the estate again this year,
high up on the slope of the high Himalayas in Dehradun's fanciest
neighborhood. One of the Indians is Dr. Raju, who was a good friend
and colleague of Ranga Rao. A servant, an old man with a wool cap,
comes to the gate in the rough brick wall that surrounds the place, and
lets us pass. A large pile of gray and purple shale lies to the side of the
house. I know that there are fossils in there—that pile is more valuable
than gold to me—but I keep walking. It looks like rain.

The house seems like a ghost. It is surrounded by verandas covered
with construction material. There are large windows of all shapes and
sizes, but it is dark inside. It appears unoccupied, and parts of it are not
finished. The lady meets us at a smaller house behind the big one. She is
a small woman, with wrinkled and yellowish skin, bent by age, her face
frozen in a scowl, unsmiling. Her unkempt gray hair is in a bun, and she
wears striped pajama pants and a flowered blouse. But you can also see
that she used to be tall, strong, and beautiful. Her eyes are piercing, light
blue, and they look straight into the heart.

We sit down for tea. She does most of the talking. She has a lot to
say. For the rest of us, it is difficult to be heard, because she is quite

deaf. She explains the injustice done to her and her husband, first by the paleontologists, then by all the Indians that have stolen her money and her things, from carpenters, to bank employees, to grocers.

Her life story is a tale of stubbornness, obsession, and sorrow. Her father was a World War I soldier, but a pacifist in World War II, shunned by the Nazis. She got married just before the German army invaded France, Belgium, and my native country, the Netherlands. Her new husband, an engineer, was in the army and died in that invasion. She was a young female student, in a country that was slipping into totalitarianism and militarism. However, that did not scare her, and she pursued her education with great effort, studying with some of the greatest minds of German science. One of her professors was Willi Hennig, the father of the modern science of systematics. Hennig believed that Germany would win the war. Never shy to have strong opinions, she challenged him, saying that Germany would lose.

"We are in the middle, a hare surrounded by foxes, we have nowhere to run," she says, quoting herself. Hennig's reply came confidently: "We are not a hare."

After choosing paleontology as her field, she worked with Professor Dehm in Munich to get a PhD degree. Dehm was on a long fossil-collection trip when the war broke out, and got stuck in Australia. He was unable to return to Germany. The authorities let him go on the condition that he sign a letter promising on his honor not to become a soldier in the German army. He did; he kept his word and stayed out of the war. The Nazis, unhappy with him, pushed him out of his job in the Nazi heartland of Bavaria to the scientific and societal backwater of Strasbourg.

After the war, with Germany shattered, Dehm moved back to Munich, and Frau Obergfell became his assistant. The allies gave her a certificate indicating that she had not been involved with the Nazis. It was there that she met Ranga Rao, some twenty years later, and married him.

She looks at me. "You cannot trust any Indians, they are all liars." I wince, but keep my mouth shut, sitting there with one crazy German lady and four Indian colleagues whom I trust and respect.

I try to change the subject, explaining that those fossils are important and that I want to study them. I ask her permission. She ignores the question, maybe she does not hear it, but I think she does. She tells us that the fossils are part of a trust that she and her late husband founded. The fossils will be prepared and studied at the trust, under her supervision. She wants to make this house a center for the study of the fossils that her husband found. She leads us on a tour of the big house, a skeletal mansion haunted by the broken hopes of the Frau and Ranga Rao. There are still shipping crates, unopened, from when she came from Europe, more than thirty years ago; there are no light fixtures or furniture. Her nephew steadies her by holding her hand as she goes up the spiral stairwell. The plans are grand: here will be the room with the fossil collection, here is the library, here the map room. Her vision for the place is a stifled obsession—the real and perceived injustices have sapped her initiative, and nothing has happened here for years.

"How long will you be here?" she asks me.

I hesitate. "We leave Dehradun tomorrow," I say, raising my voice to be heard.

"That is too short, you cannot do anything."

I do not understand. Too short for what? Is that an implicit permission for me to study the fossils? A light at the end of the tunnel?

But then the conversation goes nowhere, and it gets late. She repeats some stories, and tells more, about Sahni, Ranga Rao, and how hard her life is here. I throw in the towel and signal the others that we should leave.

It is now pouring outside. We stand on the porch of the big house, in full view of the heap of rocks. She says the collection needs to be properly displayed and housed before anyone can work on it, but that I am welcome to visit her anytime. Then comes my opening.

"The heap only contains stones from Kalakot," she says. "Maybe useful for microfossils."

"Can I take a few blocks and see if I can find some microfossils?"

"Of course you can. There are no large fossils there, only dust. We would not put fossils outside." I keep a straight face as the victory bell tolls in my head. I know that there are mammal fossils in those rocks, and this is my chance to get some. Unfortunately, it is dark and the rain is coming down in thick streams; it has turned the path to the car into a muddy stream.

"OK," I say, "we will come back tomorrow and get some blocks from the heap."

"That is no problem," she says. "What time?"

"Between nine and ten," I shout, over the rain and to penetrate her failing ears.

"I will serve you breakfast," she says.

The mood in the car is jubilant. We stop and buy beer and whiskey, ready to celebrate. Raju calls his wife, asking her to get snacks ready. Then a cellphone rings. Raju answers. The conversation is in Hindi. He puts the phone away. "Not good news," he says.

I sit in suspense while we drive, arrive, and start our party. I try to read Raju's mind but cannot. It kills me, but I have to let the Indian rules of hospitality run their course. We are his guests: he decides what to do next. Much later, with full bellies and empty glasses, he explains that the servant in the wool hat has told the Frau that there are indeed fossils in the heap, not just rocks. As a result, she has changed her mind—we are not to take any, tomorrow's meeting is canceled. The mood of the party changes, first to despair, but then to a reassessment of strategy. I want to go visit her, alone this time, and talk to her, in German. If she does not trust Indians, that is her issue. I am not Indian, she should trust me. The Indian colleagues agree.

However, I will not be able to persuade the servants to open the gate for me. We decide that Dr. Raju's wife will come with me. She is a sweet and kind woman, young looking, but with a son who is in his thirties. She also knows Frau Obergfell, and speaks English, albeit enriched with some native grammar and accents.

"She will not yield," is her assessment.

"I have 'no' now," I answer. "The worst that can happen is that I get more 'no'."

I practice what I want to tell her, four or five times, in German, avoiding words which I do not remember in that language: "to trust" and "to deceive." I drive through Dehradun for hours and eventually find a bottle of wine as a gift to mollify her.

Unannounced, Mrs. Raju and I drive to the house, riding in silence, both of us tense. She is dressed in a black and gray sari, a very impractical and beautiful garment that is basically a long stretch of fabric that is wrapped around the body using a complex set of folds, tucks, and creases. Just about any movement may result in dissolution of the ensemble, and one hand should be kept free to check such immodesty from unfolding.

The gate is locked. Mrs. Raju calls the names of the servants and we honk the horn, but no one comes. We wait for thirty minutes, and try again. No response. Mrs. Raju, age and sari notwithstanding, now climbs across the rough three-foot brick fence, steadying herself with one hand while controlling the sari with the other. She disappears toward the house. Fifteen minutes pass and she returns. A female servant has told her that madam is sick and cannot be disturbed: she has been upset and unable to sleep. We decide to give up. I will just leave her the wine and my card. Just then the servant in the wool hat comes home with fresh vegetables from the market. Mrs. Raju talks to him, and they both climb over the fence and disappear. My patience is tried for another fifteen minutes, and then the gate swings open. Battle half-won.

The house is dark, and madam is a small bundle of blankets on the couch. But feisty she remains, launching into another diatribe about people stealing from her. I sit in silence, not finding room to interject, and not knowing what to interject. Mrs. Raju takes the initiative, raising her voice to be heard. She makes the same very uncomfortable point, time and time again. "You just trust Raju and this man only," she says, comparing her husband and myself favorably to the rest of humanity. Her little finger sweeps the air, describing the lower half of a circle, a typical Indian gesture of emphasis.

The Frau is not convinced. I interject some of my points, but skip the German. "Professor Dehm gave his word of honor to leave Australia; I do, too. I am Dutch, you can trust my word of honor, even if all these Indians are bad people."

"You work with Indians, yes?" the Frau shoots at me.

"Yes, I work with Indians studying fossil whales in Gujarat, it is necessary. You were married to an Indian, yes?"

"You cannot trust them. No Indian can study these fossils. I do not understand your hurry. You have other things you study, why do you want to work on these fossils now? I will not give them up. The fossils stay."

"When you die, the Indians will take these stones and throw them into the river. For them, these are just stones—they are not paleontologists like you and me."

"Why don't you study other fossils with your Indians?"

"This is an important collection. Your husband's work is not completed. *These* fossils need to be studied."

Mrs. Raju breaks in again, adjusting the pillow to support the Frau's back. She complains loudly, but it does not stop Mrs. Raju.

"The heap will not go anywhere. You can study it later, when the trust is in effect."

"Those fossils are in the rain and sun. They are eroding. They are being destroyed."

She sags back on her side, a tear in her blue eye. I don't think it is emotion, but I don't know.

"I want to take two blocks from the heap. If there are fossils, I will extract them. I have a preparator. I will return what I find to you next year. You have my word of honor that I will return them. You have nothing to lose. All I will take is two stones, and they are worth nothing now, you have said so yourself. I will return them, and then you will see that you can trust me."

Mrs. Raju again interrupts. "You trust Tewson, he is not an Indian." My name is difficult for her, but I like the accent.

"Why do you not prepare them here?" the Frau asks, and Mrs. Raju throws me a look that asks the same question.

"It takes fancy equipment, water, electricity. It is not fast. It cannot be done here."

"I cannot guarantee water. It is often interrupted. The city does not provide."

I repeat my planned speech about my word of honor being as good as that of Dehm.

"How did you find out about this collection? Who told you? No one knows about it except the Indians."

I had foreseen that one. "I heard it first from Dr. Neil Wells, who visited Ranga Rao in Dehradun a long time ago, twenty years ago, to see the collection."

"Who?"

"Dr. Neil Wells. *Neil Wells.*"

"Who?"

"NEIL WELLS."

"Newell? I don't know a Newell."

Mrs. Raju jumps in. "Nejl Wehls." In unison, she and I chant the name, as if it were a god's name like the ones incessantly chanted in the nearby temples by revelers asking for favors: "Hare Ram, Hare Ram, Hare Ram." We, too, are praying for favors, and the slapstick aspect of this visit is not lost on me. However, right now, I must not smile. Finally, she hears and understands.

"Him I do not know, but it is possible, I may have forgotten if he has visited us."

She sits up and launches into another tirade against Indians. How the carpenter stole her stuff. How the workmen took the teacups she imported from Germany, and her Chinese porcelain. "No fossil will leave. I cannot trust anyone."

"You have my word of honor. Does my word of honor mean nothing? Are you saying I am a liar?"

Mrs. Raju goes again. "You trust two people, Raju and this man."

Her light-blue eyes suddenly look straight at me now, framed by the pale wrinkled face. "Why don't you take a sack and take some stones from the heap. Put them in your suitcase, do not show them to any

Indian, and return them when you come again. I have no objection. You I can trust."

I am shocked into silence. Somehow my own proposal has been turned around as if it were her command. I missed the switch.

"It will take time to extract the fossils. I leave for the U.S. in seven days. I will return them when I come to India again—I come every year."

"I do not doubt that you are honest, but no Indian should be involved."

"This is not a problem. I will put them in my suitcase. I will get the bag and the blocks, and come back to the house to show you what I took."

"You do not need to show me. I trust you."

I race over to the heap, which is slippery from yesterday's rain, and quickly make my choice before she changes her mind. This batch is for the principle, not the fossils. Bahadur, the Nepalese servant with the wool hat, comes and helps. He sees a stone with a tooth. He's got good eyes and seems happy to help me.

I take the bag back to the house, but she does not want to see its contents. Satisfied, and full of hope, I leave Dehradun, and then India.

THE ANCESTORS OF WHALES

The fossils are prepared, and I duly return them the next year, getting more blocks to take to the United States. Eventually jaws and thighbones, astragali and hipbones emerge from their ancient rocky prison. The vast majority are of the same species, a raccoon-sized artiodactyl called *Indohyus* which is closely related to *Khirtharia* from Pakistan. *Indohyus* was originally discovered by Ranga Rao when he found a few jaws of the animal in these rocks.[2] Most important are the skulls. We have four of them. My new preparator, Rick, is very patient with them and does a beautiful job, scraping the purple and gray sediment out of the tiniest cracks without hurting the bright white bone. I check on his progress daily, and we talk about which block to work on next. Rick has been deaf since birth, and our conversations are a mix of enunciating exaggeratedly, repeating, and pointing, with his eyes jumping from the fossils to my lips as he lip-reads. Then, one day as I walk into the prep lab, Rick apologizes for having broken a piece off one of the skulls. The break is straight, and no pieces were lost, so it is easily glued back, he says. It is not uncommon for bones to break during preparation, and as long as the break is clean, fixing them is not a big deal. As I look at the skull, I realize that the bone

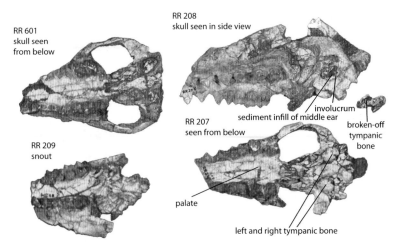

FIGURE 63. Fossil skulls of *Indohyus*. RR 208 shows the broken tympanic discussed in the text.

that broke is the tympanic. It snapped right through the middle, exposing the sediment-filled middle ear cavity. To my shock, the inside wall of the tympanic is much thicker than the outside. *Indohyus* had an involucrum, just like whales—an amazing discovery brought about by Rick's accident (figure 63). We're not gluing this one back!

Now the work takes on a frantic pace, and, in 2007, we are ready to publish. Then, in July, news reaches me that Dr. Friedlinde Obergfell has died. It is sad that she died just short of the recognition for her husband's fossils that she sought for thirty years. She leaves all her belongings to the trust that is dedicated to Ranga Rao's fossils, and to my great surprise appoints me as the main person studying the fossils. Following her wishes, she is buried on her property wearing the army jacket of her first husband and a *shalwar*, thin, loose-fitting Indian pants. I correspond with her relatives in Europe, and travel to south India to see her husband's relatives. In December of that year, we publish our work on *Indohyus*.[3]

INDOHYUS

Indohyus (figure 64) is part of a small band of artiodactyls that constitutes the family Raoellidae, named in honor of Ranga Rao by Ashok Sahni. For nearly all of them, only teeth are known, and the skull and skeleton are only known for one genus of raoellid: *Indohyus*. All known skulls and bones came from the blocks in Ranga Rao's yard in Dehradun.

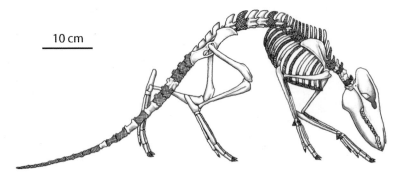

FIGURE 64. The skeleton of *Indohyus*. Parts that were not discovered are shaded. Reprinted from J. G. M. Thewissen, L. N. Cooper, M. T. Clementz, S. Bajpai, and B. N. Tiwari, "Whales Originated from Aquatic Artiodactyls in the Eocene Epoch of India," *Nature* 450 (2007): 1190–94.

These animals looked like a tiny, somewhat heavy-set deer (figure 65). Indeed, mouse deer are similar: they are very small modern artiodactyls (*Tragulus* and *Hyemoschus*) that live deep in the forests of Central Africa and Southeast Asia.

Raoellids are only known from South Asia, Pakistan, and India (figure 22)—there is a questionable record from Myanmar.[4] The oldest raoellids are from the Chorgali Formation of Pakistan, around fifty-two million years ago, and the youngest are probably from Kalakot, approximately forty-six million years old. The family is not a very coherent group. Scientists have referred new fossil specimens to the group without studying all those that were already known; as a result, the group has become a somewhat chaotic assemblage. It would be useful for someone to study the entire group carefully. But such a systematic revision will not be easy. Many species are known only from a few teeth, and the fossils are dispersed over three continents in about a dozen labs and museums. Skulls and skeletons are only known for *Indohyus*; for the other genera, such as *Khirtharia* and *Kunmunella,* mostly teeth are known, and it is possible that some really belong in different artiodactyl families.

The thickened lip of the tympanic bone, the involucrum, gave us a clue that *Indohyus* is closely related to whales, but that idea has to be more formally investigated. Our cladistic analysis (see chapter 10) shows that whales were indeed more closely related to *Indohyus* than to any other artiodactyl, including hippos. Later it was shown that, in addition, hippos are the closest relatives of the raoellid-cetacean group (figure 66).[5] That result actually does not conflict with the molecular

FIGURE 65. Life reconstruction of the Eocene artiodactyl *Indohyus*, the closest extinct relative of whales. *Indohyus* is in the family Raoellidae, which lived in South Asia from forty-six to fifty-two million years ago.

data that show that hippos and whales are the closest relatives, because the molecular studies could not include fossil animals. In other words, hippos may still be the closest *living* relative, it is just that extinct *Indohyus* is even closer. In addition to the presence of an involucrum, these groups share a number of dental features, such as the front-to-back arrangement of the upper incisors in the jaw, and the high triangular

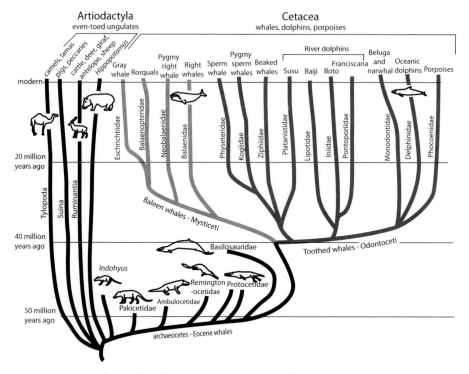

FIGURE 66. Relationships of cetaceans to artiodactyls. All groups of modern artiodactyls and cetaceans are included in this figure. Not shown are a large number of extinct groups that are not discussed in this book.

crowns of the posterior premolars. Tooth-wear patterns of *Indohyus* show the same specializations as cetaceans (figure 50).

With that, it seems that the issue of cetacean relations is finally resolved. Mesonychians are not related to cetaceans. The fossil evidence shows that cetaceans are derived from a basal, Eocene artiodactyl, and the closest modern relative of cetaceans is the hippo.

However, that is not the end of the story. Yet another cladistics analysis,[6] on a slightly different data-set, confirmed most of the results just discussed, but found that the support for that view is only slightly stronger than that for the old mesonychian idea. It is as if those fierce predators are still waiting in the wings to reclaim their position next to the majestic whales by pouncing on little vegetarian *Indohyus*. As one well-known mammalogist put it: systematics is the soap opera of biology.

With all the rearrangements of the relationships of whales, we have to wonder whether it would be useful to actually include *Indohyus* within Cetacea instead of keeping it just outside the group. After all, there is nothing sacred about having *Pakicetus* be the basal cetacean. And the main feature that characterizes cetaceans, the involucrum, also occurs in *Indohyus*. Furthermore, if names should be used for groups that include an ancestor and all its descendants (monophyletic groups), the term *artiodactyl* should now include all cetaceans as well, since they too descended from the ancestral artiodactyl.

Some authors have indeed advocated for changing the meanings of Cetacea and Artiodactyla in one way or another,[7] but I do not agree with them. The term *artiodactyl* has been around for more than 150 years and has had a stable and biologically coherent meaning. Changing it now would only cause confusion, especially if not all authors follow the same meaning. It would be especially bad for new students, who would quickly get confused by names that differ in meaning depending on when and by whom they are used. My preference is to stick with the old meaning of the word, where Artiodactyla does not include Cetacea, and accept the fact that the former does not include all descendants of that first ancestor. Scientists call this a paraphyletic group, and Artiodactyla would be one of those.

Similarly, Cetacea for decades has meant *Pakicetus* and all its descendants. It too is a biologically coherent group of aquatic predators, albeit that some had legs and walked. Adding *Indohyus* to this group muddies the water: it is so different biologically that it would render the term Cetacea meaningless in any sense *except* systematically.

Feeding and Diet. In general, *Indohyus* has very typical artiodactyl teeth: a dental formula of 3.1.4.3/3.1.4.3, with upper molars that bear four cusps, and lower molars that have a high trigonid, with two cusps, and a low talonid, also with two cusps (figure 34). The shape of these cusps differs among raoellids: in *Indohyus* and *Kunmunella,* the cusps are sharp and connected by a weak crest, whereas in *Khirtharia*, the cusps are low and blunt. Among modern mammals, the first molar type is common in leaf eaters, while the latter is common in fruit eaters, but it is not clear whether this difference holds for raoellids. Without doubt, those dental differences relate to diet and food processing somehow, but it is unclear how. Stable carbon isotope data indicate that *Indohyus* and *Khirtharia* both fed on land plants.[8]

Another clue to food processing comes from the relative position of the joint between lower jaw and skull, where the skull has a socket in which a ball-like joint, the mandibular condyle, fits. As shown in figure

25, the condyle is well above the level of the tooth row in herbivores such as deer. In meat eaters, such as the whales shown in this figure, the condyle is at the same height as the tooth row. *Indohyus's* jaw has an herbivore's shape, as expected.

While the molars of *Indohyus* are not very specialized, their tooth wear is. Early whales are characterized by nearly exclusive phase I wear on their lower molars (figure 50, see chapter 11). Eocene artiodactyls show a combination of phase I, phase II, and apical wear. In *Indohyus*, all three wear types are present, but phase I dominates. Apparently, the land plants that *Indohyus* ate were processed in ways different from the way other Eocene artiodactyls processed their food. The dentition also provides other clues to feeding. *Indohyus* had a long and pointed snout, with incisors arranged from front to back, not side to side. This may have been a specialized mechanism for cropping certain plants. In addition, its premolars have high crowns with sharp cutting edges on their sides. At present, the function of these features is not understood, but with the hundreds of fossils of *Indohyus* that are known from Kalakot, and a few more years of study, I am very hopeful that we will know.

Vision and Hearing. The eyes of *Indohyus* are located on the side of the skull, as is common in land mammals, and unlike just about all fossil whales (figure 52). This part of the skull is highly variable and highly specialized in all Eocene whales, and *Indohyus* lacks these specializations. The distance between *Indohyus's* orbits and its brain, the intertemporal area, is similar to other artiodactyls, and unlike Eocene whales.

The forces related to the continental collision between Asia and India millions of years ago actually affect how well we can study *Indohyus* in the present. Mountain building deformed the rocks and their fossils, flattening skulls and breaking bones. The skulls of the animal were crushed, and delicate structures were obliterated. Except for the presence of an involucrum, very little is known about its ear.

Walking and Swimming. Overall, the skeleton of *Indohyus* resembles that of unspecialized artiodactyls, adapted for land locomotion, and with some specializations often found in runners.[9] There are five fingers and four or five toes, and *Indohyus* was digitigrade, like a dog, not like its artiodactyl relatives, who walk on the tips of their toes using hooves (unguligrade).

In spite of this, there are two lines of evidence that indicate that *Indohyus* was not a fully terrestrial species. First, some of the bones of *Indo-*

hyus have a thick outside layer, the cortex, suggesting that one of their functions is to be ballast while the animal is in the water (figure 62). That resembles pakicetids, who show this tendency to osteosclerosis to a greater degree. Also, the oxygen isotopes are interesting. In chapter 9, isotopes were used to investigate the source of drinking water for some of the early whales, but here they can help us with another problem. The ratio of ^{18}O and ^{16}O in the water inside the body of an animal is reflected in its bones and teeth. Animals lose body water in a number of different ways, for instance when they pee, and for females, when they produce milk. They also lose body water when it evaporates through the skin. Interestingly, evaporation through the skin is a process in which the isotopes are fractionated: water with the lighter oxygen isotope is more likely to go into the gas phase and disappear from the body than water with the heavier isotope. As a result, animals that lose a lot of body water through their skin have an isotopic signature that is skewed toward the heavier isotope. Mammals that live in water do not sweat or evaporate water, so isotope ratios can help to discern whether they are aquatic. Indeed, oxygen-isotope values for *Indohyus* indicate that it spent time in water.

Habitat and Ecology. *Indohyus* presents a paradox. On the one hand, carbon-isotope values and the molar shape suggest life on land; on the other hand, oxygen isotopes and osteosclerosis suggest freshwater. A possible resolution for the paradox comes from studying a modern mammal: the mouse deer. Mouse deer live on land, eating the flowers, leaves, and fruits of terrestrial plants. However, they are always found near rivers, and when in danger, mouse deer jump into the water and hide.[10] Mouse deer bones are not osteosclerotic, and *Indohyus* is not closely related to them. However, they may be the perfect ecological equivalent. Here, then, may be the key to the origin of aquatic life for whales: predator-avoidance behavior in their early artiodactyl ancestors.

Kalakot, the fossil site where *Indohyus* is abundant, has not been studied sedimentologically, and not much is known about the habitat these animals were living in. What is known is that there must have been hundreds of skeletons of *Indohyus* all washed together buried and mixed with just a few other forms. Some of the bones found here are articulated, but most are not. Apparently, there was time for rotting to disarticulate many of the skeletons. It is possible that this was a floodplain of a river with animals living and dying. Skeletons accumulated on the plain, were dispersed by scavengers, and during the next flood, washed together into streams.

A TRUST FOR FOSSILS

The pile of rocks holding *Indohyus* still sits on the estate on Rajpur Road, with the grave of Friedlinde Obergfell nearby, guarded by Bahadur and his wife. The extracted fossils are now safe in the unfinished house, and we have started to sort through the bags of fossils in the cellar. More fossils are being extracted every day, but the work is slow, and there is no money to hire a fossil preparator. The trust that manages the *Indohyus* fossils has a home, and fossils, and a mission to study them, but there are no funds to get the study off the ground and save the *Indohyus* fossils for the future. I hope to avoid it, but it is possible that the entire place will fold. Sadly, that may be a fitting end to the tragic tale of Friedlinde Obergfell and Anne Ranga Rao.

CHAPTER 15

The Way Forward

THE BIG QUESTION

I love to talk about whale evolution, and my audiences range from fifth graders, to our local Rotary club, to cetologists at international meetings. To point out how dramatic the evolution of whales is, I usually start by asking people to think about two fancy vehicles. I could use a bullet train and a nuclear submarine, but, because it is less intimidating, I ask them to think about the Batmobile and the Beatles' Yellow Submarine. Whales started out with a very elaborately perfected body adapted to life on land. They changed it, in about eight million years, to a body perfectly tuned to the ocean. I ask the audience to imagine getting a team of engineers together to take the Batmobile apart and build the Yellow Submarine from its parts. Just about everything that works well on land will fail miserably in water. All the organ systems have to change—from locomotion, to sense organs, to osmoregulation, to reproduction. And of course, in evolution, all the intermediate species were functional in their environment. Adding that requirement would dictate that at the end of every working day our engineers can still present a working vehicle. It would be an impossible job, and that indicates how remarkable a transition this really was. And now, remarkably, it is all documented by fossils.

After such public lectures, the question I am asked most commonly is *why* cetaceans went into the water. There is much that we still do not

208 | Chapter 15

know, but if I step back from all the details and squint my eyes, a blurry movie reel becomes visible (figure 66). Little raccoon-sized artiodactyls were eating flowers and leaves, but hid in the water from danger. Their descendants stayed there, now hiding in the water as predators, spying for prey. *Their* descendants learned how to swim fast, pursued new prey, and little by little, lost the ability to get around on land. After experimenting with different ways of swimming, they eventually changed their bodies to be sleek and streamlined. Thus all ties to the land were broken. One group added a sound-emission system to its already highly developed hearing system in order to locate prey: the echolocating odontocetes or toothed whales. The other group evolved baleen, used for grazing the krill fields: the mysticetes or baleen whales. There was no single drive to get from land to water. Cetaceans took small steps, not in a straight line, and most related somehow to feeding and diet. Each of those steps was opportunistic, and there were plenty of failed experiments.

In spite of what we know already, there are lots of interesting questions left, but one strikes me as the Big Question. Mammals, in general, are highly integrated and built on a more constrained blueprint than groups such as reptiles and fish. For instance, the dental formula for placental mammals is 3.1.4.3/3.1.4.3 and hardly ever goes up in numbers (see figure 11 and chapter 2). Also, in mammals, there are at most three phalanges per finger and two for the thumb (figure 13), and there are around twenty-six vertebrae in front of the sacrum (chapter 12). In all of those design features, mammals are more constrained than fish, amphibians, and reptiles. But cetaceans are the exception. They make a mockery of the mammalian rules, varying wildly in numbers of teeth, numbers of phalanges, and numbers of presacral vertebrae. It is as if some very basic mammalian rules governing development have been broken. And paradoxically, in spite of that greater-than-normal variation, all modern cetaceans look quite similar on the outside. They all have a streamlined body, are basically naked, lack a neck, have flippers for forelimbs, got rid of their hind limbs, and evolved a horizontal fluke for a tail. If they are built on a relaxed blueprint, why are they so similar externally? My Big Question is about the genetic switches that caused the blueprint to relax, and how they affect the paradox of the conservative external morphology.

The first part of that question is already so broad that it cannot be phrased as an explicit hypothesis that is testable in the way most sciences operate. Instead, we need to break it up into smaller, more specific and answerable questions. Some of those answers will come from fossils.

Only fossils can show us what actually happened in evolution. But many of the answers will come from studying the genes that channel the development of an embryo. Those can only be studied in modern whales.

In our quest for answerable questions, we should start with just one organ system and make sense of what developmental data and paleontological data have to offer together. Given that feeding evolution is central to early whale evolution, it makes sense to start there.

TOOTH DEVELOPMENT

With it being as difficult as it is to get cetacean embryos, the most straightforward way to start this project is to go with the dolphin embryos that John Heyning and Bill Perrin gave me, years ago, when I was studying hind-limb loss. Those embryos are all from one species, the pantropical spotted dolphin, *Stenella attenuata*. As an adult, this species has more than thirty-five teeth in each upper and lower jaw, way more than the eleven its Eocene ancestors had. The teeth are tiny, as shown for a related dolphin in figure 25. More than eleven teeth per half jaw is called polydonty. Polydonty only occurs in two groups of mammals besides cetaceans: manatees and the giant armadillo *Priodontes*. In addition to being polydont, *Stenella* is also homodont: all its teeth look the same. There is no distinction between incisors, canine, premolars, and molars. Homodonty occurs, to some degree, in all modern cetaceans that still have teeth, but is rare in other mammals. Interestingly, the Eocene cetaceans that I study are neither homodont nor polydont. Both features show up gradually but more or less simultaneously, starting around thirty-four million years ago. Early baleen whales still have teeth; there are fifteen to twenty teeth per jaw. Those teeth are more similar than those of Eocene whales, but there is a definite tendency toward homodonty. The same happens, independently, in early odontocetes. That makes me think that there is a relation between homodonty and polydonty.

We know a lot about tooth development from biomedical studies, mostly on mice. When the embryo is still tiny, and long before there are any teeth, a protein is made in the front of the jaw that goes by the acronym BMP4. Another protein, FGF8, is made in the back of the jaw.[1] Interestingly, in other vertebrates, BMP4 occurs throughout the jaw, and FGF8 is not involved in tooth development at this stage.[2] And, of course, those other vertebrates are homodont, or nearly so. Experiments have been done with mouse embryos, and, sure enough, if the embryo is

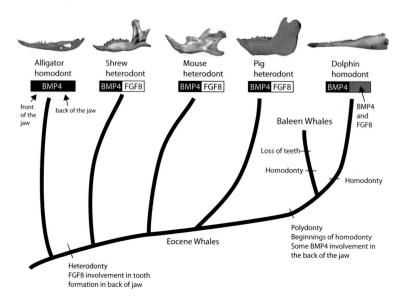

FIGURE 67. Genes determining tooth shape make the proteins BMP4 (black bar) and FGF8 (white bar). These proteins both occur in the jaw of different vertebrates, and the pattern varies between reptiles (alligator) and mammals (all others). The pattern of these proteins in dolphins is different from that of the other mammals. Alligators and dolphins have similar teeth across the tooth row (homodonty), but their dental shape results from different gene expression patterns. The branching diagram at the bottom summarizes evolutionary events leading to the changes in protein distribution. The Eocene whales of this book are on the segment to baleen whales and dolphin.

tricked into making BMP4 in the back of the jaw too, the mouse's molars become simpler, and all the teeth look like incisors.[3] It turns out that, in dolphin embryos, BMP4 is still present in the front and FGF8 in the back, but the back of the jaw also has BMP4.[4] It appears, then, that the interaction between these two proteins within the jaw could be part of an important evolutionary switch: FGF8 taking a role in tooth development is a novelty for mammals, and the expansion of BMP4 overprints that role in cetaceans (figure 67). Also relevant is that the presence of these proteins in the embryo actually occurs long before there are morphological signs of teeth, while the morphological result (homodonty or heterodonty) can only be seen when the teeth are formed, much later in development. That may make it unlikely that those two patterns, polydonty and homodonty, are the result of one simple genetic event at a single time in development, although we can't be sure about that at this point.

Our study on dolphin tooth development was a good first step toward understanding the shapes of teeth and the genes leading to those shapes.[5] It said something about homodonty but did not find a direct mechanism that linked homodonty to polydonty, and only involved a single species of toothed whale. I need embryos for more species, specifically ones not closely related to dolphins, like baleen whales. Those are even harder to get than dolphin embryos.

BALEEN AS TEETH

Baleen whales do not have teeth, but their embryos do.[6] Just as with the hind limb buds, teeth are formed in the jaws of the tiny embryos, and later in development, these tooth precursors cease to grow, and languish. These little tooth buds even grow into tiny mineralized structures in some baleen whales,[7] but in no baleen whale do they ever erupt from the gums. Near the time that the teeth disappear, baleen starts to develop from the same area of the upper jaw where the teeth used to be.[8] The similarity in timing makes it tempting to speculate that baleen formation is somehow linked to the cessation of tooth formation. The fossil record gives some clues, too. Baleen does not fossilize, but it has been suggested that the presence of baleen in a fossil whale can be deduced from grooves on the palate.[9] These grooves carry blood vessels, and a fast-growing tissue like baleen needs a lot of blood to supply it. Following this logic, it has been suggested that some Oligocene mysticetes had the beginnings of baleen formation even though they still had teeth. In fact, the dentition of those whales was polydont and, to a large extent, homodont. Similarly, there are many baleen plates, and they are all very similar.

Baleen forms as a thickening of the epithelium of the upper jaw, and interestingly, teeth initially also form as a thickening. In the case of teeth, the thickening buries itself into the underlying tissue, the mesenchyme. If the two processes are linked, I would expect that a subset of the genes involved in tooth formation is also involved in baleen formation. Genes that often work together in building different organs are referred to as genetic toolkits. It is possible that early in baleen whale evolution the toolkit that built teeth was reprogrammed to build baleen instead. That shift of the toolkit caused the teeth to disappear. Such a novel function for an existing process has been called exaptation. If I can show that the same genetic toolkit is involved in both tooth loss and baleen formation, the next question will be whether that toolkit also operates in other processes—the lack of hair development, for instance, since hair development also has similarities with tooth development in the embryo. If that is true, it is possible that a few changes in a key group of regulatory genes would affect a whole array of cetacean organs and drive the evolution of the group.

I think about this as I fly to Barrow, on Alaska's North Slope. There I hope to study embryos of the bowhead whale, a species of baleen whale, that have been harvested by Iñupiat eskimos. Circumarctic indigenous people have subsisted on bowhead for many centuries, and the International Whaling Commission strictly regulates this, so that it does not affect the health of the bowhead populations, which are growing right now. I am years away from answering even the simplest questions related to bowhead development, but it makes me think back to that first trip to Pakistan, in 1991. I did not go to Pakistan to answer the questions that I ended up answering, and the outbreak of the war all but killed my first field season. Only by perseverance and luck was I able to follow through, and that field season paved the way for the exciting discoveries that I was part of later. Another decade will be needed to show how much cetacean embryos can enhance the fossil story.

For now, I am content that this book has summarized the remarkable progress that has been made in our understanding of whale origins. The subject used to be undocumented, hard to grasp conceptually, and the darling of creationists for its absence of fossils. Now it is the darling of evolutionary biology textbooks: it is well understood, with plenty of intermediate fossils, many clear-cut functional links, and the beginnings of an understanding of the molecular mechanisms that drive it all. Many questions remain, and without doubt, pieces of this story will have to be rewritten as we learn more. But that is part of the normal dynamics of science. New finds are used to test past conclusions, and with every step we get closer to true understanding. It is also part of the normal dynamics of human life. With every experience that a human has, growth occurs and old ideas are resculpted. For whale origins, amazing things have been learned over the past two decades, and I hope a new generation of budding scientists will push our understanding of whale evolution beyond our present horizon. It is your turn—go for it.

Notes

CHAPTER I. A WASTED DIG

1. R.M. West, "Middle Eocene Large Mammal Assemblage with Tethyan Affinities, Ganda Kas Region, Pakistan," *Journal of Paleontology* 54 (1980): 508–33.

2. P.D. Gingerich and D.E. Russell, "*Pakicetus inachus,* a New Archaeocete (Mammalia, Cetacea)," *Contributions from the Museum of Paleontology, University of Michigan* 25 (1981): 235–46. P.D. Gingerich, N.A. Wells, D.E. Russell, and S. M. I. Shah, "Origin of Whales in Epicontinental Remnant Seas: New Evidence from the Early Eocene of Pakistan," *Science* 220 (1983): 403–06.

3. D.T. Gish, *Evolution: The Challenge of the Fossil Record* (El Cajon, CA: Creation-Life Publishers, 1985).

4. A. Boyden and D. Gemeroy, "The Relative Position of the Cetacea among Orders of Mammalia as Indicated by Precipitin Tests," *Zoologica* 35 (1950): 145–51. M. Goodman, J. Czelusniak, and J.E. Beeber, "Phylogeny of Primates and Other Eutherian Orders: A Cladistics Analysis Using Amino Acid and Nucleotide Sequence Data," *Cladistics* 1 (1985): 171–85.

5. D. Gish, "When Is a Whale a Whale?" *Acts & Facts* 23 (1994, No. 4). http://www.icr.org/article/when-whale-whale/.

6. Different scientists use the word *whale* differently. In this book, for fossil species, *whale* and *cetacean* are used interchangeably. As such, *whales* includes fossil dolphins and porpoises.

7. J.G.M. Thewissen and S.T. Hussain, 1993, "Origin of Underwater Hearing in Whales," *Nature* 361 (1993): 444–45.

CHAPTER 2. FISH, MAMMAL, OR DINOSAUR?

1. Aristotle, *Historia Animalium*, Book III, http://web.archive.org/web/20110215182616/http://etext.lib.virginia.edu/etcbin/toccer-new2?id=

AriHian.xml&images=images/modeng&data=/texts/english/modeng/parsed&tag
=public&part=3&division=div2.

2. There are some cetaceans that pertain to the Odontoceti that have barely any teeth. A male narwhal has only one, a tusk longer than the animal, whereas a female narwhal has no teeth that break through the gums at all, and the same is true for many female beaked whales. Alternatively, some whales with teeth are not toothed whales, such as the whales that lived between 50 and 37 million years ago. The use of the phrase "toothed whales" here means odontocete.

3. D. W. Rice, *Marine Mammals of the World, Systematics and Distribution*, Special Publication Number 4 (1998), Society for Marine Mammalogy.

4. The first part of the Latin here means "a penis that enters the female, and breast that gives milk." Indeed, feeding its young with mother's milk is the critical feature for a mammal, but a male copulatory organ is not; a penis is also present in crocodiles and turtles, for instance. The last part of the quote was translated for me by Dr. Graham Burnett as "from (the authority of) the law of nature, by right and by merit," and surely exemplifies another of Melville's mischievous moments in writing this book.

5. H. Melville, *Moby-Dick; or, The Whale* (New York: Random House, 1992), 193–94.

6. C. Darwin, *The Origin of Species by Means of Natural Selection or the Preservation of Favoured Races in the Struggle for Life* (Harmondsworth: Penguin, 1968), 215.

7. Quoted in S. J. Gould, "Hooking Leviathan by Its Past," *Natural History*, May 1994: 8–15.

8. R. Harlan, "Notice of the Fossil Bones Found in the Tertiary Formation of the State of Louisiana," *Transactions of the American Philosophical Society*, N.S. 4 (1834): 397–403, pl. 20.

9. R. Owen, "Observations on the *Basilosaurus* of Dr. Harlan (*Zeuglodon cetoides*, Owen)," *Transactions of the Geological Society of London*, Ser. 2, No. 6 (1839): 69–79, pl. 7–9. R. Owen, "Observations on the Teeth of the *Zeuglodon, Basilosaurus* of Dr. Harlan," *Proceedings of the Geological Society of London* 3 (1839): 24–28.

10. International Code for Zoological Nomenclature—see http://www.nhm. ac.uk/hosted-sites/iczn/code/.

11. J. G. Wood, "The Trail of the Sea-Serpent," *Atlantic Monthly* 53 (June 1884): 799–814.

12. D. E. Jones, "Doctor Koch and his 'Immense Antediluvian Monsters,'" *Alabama Heritage* 12 (Spring 1989): 2–19, http://www.alabamaheritage.com/ vault/monsters.htm.

13. Quoted in J. D. Dana, "On Dr. Koch's Evidence with Regard to the Contemporaneity of Man and the Mastodon in Missouri, *American Journal of Science and Arts* 9 (35, 1875): 335–46.

14. J. Müller, *Über die fossilen Reste der Zeuglodonten von Nordamerica, mit Rücksicht auf die europäischen Reste dieser Familie* (Berlin: G. Reimer, 1849).

15. *Dallas Gazette* of Cahawba, Alabama, March 30, 1855, quoted in note 12.

16. P. D. Gingerich, B. H. Smith, and E. L. Simons, "Hind Limbs of Eocene *Basilosaurus*: Evidence of Feet in Whales," *Science* 229 (1990): 154–57.

17. J. Gatesy and M.A. O'Leary, "Deciphering Whale Origins with Molecules and Fossils," *Trends in Ecology & Evolution* 16 (2001): 562–70.

18. Groups of related species are included in one genus, and groups of related genera are included in one family. The most common levels of hierarchy in zoological nomenclature are: species, genus, family, superfamily, suborder, order, class, and phylum. Cetacea (cetaceans in English) is the name of an order in the class Mammalia (mammals in English). See also page 14.

19. Basilosaurines include *Basilosaurus, Chrysocetus, Cynthiacetus,* and *Basilotritus* and are found in Europe, Africa, and the Americas. Among the dorudontines, *Dorudon, Saghacetus, Masracetus,* and *Stromerius* are known from Egypt only; *Zygorhiza* lived in North America, Antarctica, and New Zealand; and *Ocucajea* and *Supayacetus* are known from Peru only.

20. M.D. Uhen, "Form, Function, and Anatomy of *Dorudon atrox* (Mammalia, Cetacea): An Archaeocete from the Middle to Late Eocene of Egypt," *University of Michigan Papers on Paleontology* 34 (2004): 1–222. This work comprehensively treats one of the best-known basilosaurids, and covers many of the topics discussed here. Citations of this and other papers of ubiquitous importance that were already cited are not repeated.

21. The third molar in the upper and lower jaw is the wisdom tooth. That tooth is present in some people, but never erupts in others.

22. R. Kellogg, *A Review of the Archaeoceti* (Washington, DC: Carnegie Institute of Washington, 1936).

23. C.C. Swift and L.G. Barnes, "Stomach Contents of *Basilosaurus Cetoides:* Implications for the Evolution of Cetacean Feeding Behavior, and Evidence for Vertebrate Fauna and Epicontinental Eocene Seas," *Abstracts of Papers, Sixth North American Paleontological Convention* (Washington, DC, 1996).

24. J.M. Fahlke, K.A. Bastl, G. Semprebon, and P.D. Gingerich, "Paleoecology of Archaeocete Whales throughout the Eocene: Dietary Adaptations Revealed by Microwear Analysis," *Palaeogeography, Palaeoclimatology, Palaeoecology* 386 (2013): 690–701. doi:10.1016/j.palaeo.2013.06.032.

25. J.M. Fahlke, "Bite Marks Revisited: Evidence for Middle-to-Late Eocene *Basilosaurus isis* Predation on *Dorudon atrox* (Both Cetacea, Basilosauridae)," *Palaeontologia Electronica* 15 (2012): 32A.

26. R.A. Dart, "The Brain of the Zeuglodontidae (Cetacea)," *Proceedings of the Zoological Society, London* 42 (1923): 615–54.

27. L. Marino, "Brain Size Evolution," in *Encyclopedia of Marine Mammals* (2nd ed.), ed. W.F. Perrin, B. Würsig, and J.G.M. Thewissen (San Diego, CA: Academic Press, 2009), 149–52.

28. T. Edinger, "Evolution of the Horse Brain," *Geological Society of America, Memoir* 25 (1948).

29. L. Marino, M.D. Uhen, B. Frohlich, J.M. Aldag, C. Blane, D. Bohaska, and F.C. Whitmore, Jr., "Endocranial Volume of Mid-Late Eocene Archaeocetes (Order: Cetacea) Revealed by Computed Tomography: Implications for Cetacean Brain Evolution," *Journal of Mammalian Evolution* 7 (2000): 81–94. L. Marino, "What Can Dolphins Tell Us about Primate Evolution?" *Evolutionary Anthropology* 5 (1997, no. 3): 81–85.

30. J.G.M. Thewissen, J. George, C. Rosa, and T. Kishida, "Olfaction and Brain Size in the Bowhead Whale," *Marine Mammal Science* 27 (2011): 282–94.

31. H.J. Jerison, *Evolution of the Brain and Intelligence* (New York: Academic Press, 1973). L. Marino, D.W. McShea, and M.D. Uhen, "Origin and Evolution of Large Brains in Toothed Whales," *Anatomical Record* 281A (2004): 1247–55. Encephalization quotient is defined as brain-weight-in-grams/0.12 body-weight-in-grams$^{0.67}$.

32. Bowhead whale 08B11 had a brain size of 2,950 grams and weighed 14,222,000 grams; see note 30.

33. W.C. Lancaster, "The Middle Ear of the Archaeoceti," *Journal of Vertebrate Paleontology* 10 (1990): 117–27.

34. V. de Buffrénil, A. de Ricqlès, C.E. Ray, and D.P. Domning, "Bone Histology of the Ribs of the Archaeocetes (Mammalia, Cetacea)," *Journal of Vertebrate Paleontology* 10 (1990): 455–66.

35. M. Taylor, "Stone, Bone, or Blubber? Buoyancy Control Strategies in Aquatic Tetrapods," in *Mechanics and Physiology of Animal Swimming*, ed. L. Maddock, Q. Bone, and J. M. V. Rayner (Cambridge: Cambridge University Press, 1994), 205–29.

36. S.I. Madar, "Structural Adaptations of Early Archeocete Long Bones," in *The Emergence of Whales*, ed. J. G. M. Thewissen (New York: Plenum Press, 1998), 353–78.

37. M. M. Moran, S. Bajpai, J. C. George, R. Suydam, S. Usip, and J. G. M. Thewissen, "Intervertebral and Epiphyseal Fusion in the Postnatal Ontogeny of Cetaceans and Terrestrial Mammals," *Journal of Mammalian Evolution* (2014), doi:10.1007/s10914-014-9256-7. M. D. Uhen, "New Material of *Natchitochia jonesi* and a Comparison of the Innominata and Locomotor Capabilities of Protocetidae," *Marine Mammal Science* (2014), doi:10.1111/mms.12100.

38. In anatomical language, the bony pelvis includes the unpaired sacrum plus the paired innominate. The innominate is also called the os coxae and is composed of ilium, ischium, and pubis. In this book, the more common English-language use of *pelvis* is followed, as a synonym of *innominate*.

39. E.A. Buchholtz, "Implications of Vertebral Morphology for Locomotor Evolution in Early Cetacea," in *The Emergence of Whales*, ed. J. G. M. Thewissen (New York: Plenum Press, 1998), 325–52.

40. F.E. Fish, "Biomechanical Perspective on the Origin of Cetacean Flukes," in *The Emergence of Whales*, ed. J. G. M. Thewissen (New York: Plenum Press, 1998), 303–24.

41. P.W. Webb and R.W. Blake, "Swimming," in *Functional Vertebrate Morphology*, ed. M. Hildebrand, D. M. Bramble, K. F. Liem, and D. B. Wake (Cambridge, MA: Harvard University Press, 1985), 110–28.

42. H. Benke, "Investigations on the Osteology and the Functional Morphology of the Flipper of Whales and Dolphins (Cetacea)," *Investigations on Cetacea* 24 (1993): 9–252.

43. L.N. Cooper, S.D. Dawson, J.S. Reidenberg, and A. Berta, "Neuromuscular Anatomy and Evolution of the Cetacean Forelimb," *Anatomical Record* 290 (2007): 1121–37.

44. J.G.M. Thewissen, L.N. Cooper, J.C. George, and S. Bajpai, "From Land to Water: The Origin of Whales, Dolphins, and Porpoises," *Evolution: Education and Outreach* 2 (2009): 272–88.

45. L. Bejder and B.K. Hall, "Limbs in Whales and Limblessness in Other Vertebrates: Mechanisms of Evolutionary and Developmental Transformation and Loss," *Evolution & Development* 4 (2002): 445–58.

46. M.D. Struthers, "The Bones, Articulations, and Muscles of the Rudimentary Hind-Limb of the Greenland Right Whale (*Balaena mysticetus*)," *Journal of Anatomy and Physiology* 15 (1881): 142–321. M.D. Struthers, 1893, "On the Rudimentary Hind Limb of the Great Fin-Whale (*Balaenoptera musculus*) in Comparison with Those of the Humpback Whale and the Greenland Right Whale," *Journal of Anatomy and Physiology* 27 (1893): 291–335.

47. F.A. Lucas, "The Pelvic Girdle of Zeuglodon, *Basilosaurus cetoides* (Owen), with Notes on Other Portions of the Skeleton," *Proceedings of the United States National Museum* 23 (1900): 327–31.

48. P.D. Gingerich, "Marine Mammals (Cetacea and Sirenia) from the Eocene of Gebel Mokattam and Fayum, Egypt: Stratigraphy, Age, and Paleoenvironments," *University of Michigan Papers on Paleontology* 30 (1992): 1–84.

49. J. Zachos, M. Pagani, L. Sloan, E. Thomas, and K. Billups, "Trends, Rhythms, and Aberrations in Global Climate 65 Ma to Present," *Science* 292 (2001): 686–93.

50. A. Haywood, *Creation and Evolution* (London: Triangle Books, 1985), quoted in note 7.

CHAPTER 3. A WHALE WITH LEGS

1. D.P. Domning and V. de Buffrénil, "Hydrostasis in the Sirenia: Quantitative Data and Functional Interpretations," *Marine Mammal Science* 7 (1991): 331–68.

2. N.A. Wells, "Transient Streams in Sand-Poor Redbeds: Early-Middle Eocene Kuldana Formation of Northern Pakistan," *Special Publication, International Association for Sedimentology,* 6 (1983): 393–403. A. Aslan and J.G.M. Thewissen, "Preliminary Evaluation of Paleosols and Implications for Interpreting Vertebrate Fossil Assemblages, Kuldana Formation, Northen Pakistan," *Palaeovertebrata* 25 (1996): 261–77.

3. R.M. West, "Middle Eocene Large Mammal Assemblage with Tethyan Affinities, Ganda Kas Region, Pakistan," *Journal of Paleontology* 54 (1980): 508–33.

4. P.D. Gingerich and D.E. Russell, "*Pakicetus inachus,* a New Archaeocete (Mammalia, Cetacea)," *Contributions from the Museum of Paleontology, University of Michigan* 25 (1981): 235–46. P.D. Gingerich, N.A. Wells, D.E. Russell, and S.M.I. Shah, "Origin of Whales in Epicontinental Remnant Seas: New Evidence from the Early Eocene of Pakistan," *Science* 220 (1983): 403–406.

5. For cetaceans, *bulla* is a synonym of *tympanic* (see chapter 1 and figure 2).

6. J.G.M. Thewissen, S.T. Hussain, and M. Arif, "Fossil Evidence for the Origin of Aquatic Locomotion in Archaeocete Whales," *Science* 263 (1994): 210–12.

7. S.J. Gould, "Hooking Leviathan by Its Past," *Natural History*, May 1994: 8–15.

CHAPTER 4. LEARNING TO SWIM

1. S.J. Gould, "Hooking Leviathan by Its Past," *Natural History*, May 1994: 8–15.

2. A.B. Howell, *Aquatic Mammals: Their Adaptations to Life in the Water* (Baltimore, MD: C.C. Thomas, 1930).

3. J.E. King, *Seals of the World* (Ithaca, NY: Cornell University Press, 1983).

4. F.E. Fish, "Function of the Compressed Tail of Surface Swimming Muskrats (*Ondatra zibethicus*)," *Journal of Mammalogy* 63 (1982): 591–97. F.E. Fish, "Mechanics, Power Output, and Efficiency of the Swimming Muskrat (*Ondatra zibethicus*)," *Journal of Experimental Biology* 110 (1984): 183–210.

5. F.E. Fish, "Dolphin Swimming: A Review," *Mammal Review* 4 (1991): 181–95. F.E. Fish, "Power Output and Propulsive Efficiency of Swimming Bottlenose Dolphins (*Tursiops truncatus*)," *Journal of Experimental Biology* 185 (1993): 179–93.

6. U.M. Norberg, "Flying, Gliding, Soaring," in *Functional Vertebrate Morphology*, ed. M. Hildebrand, D.M. Bramble, K.F. Liem, and D.B. Wake (Cambridge, MA: Belknap Press, 1985), 129–58.

7. P.W. Webb and R.W. Blake, "Swimming," in *Functional Vertebrate Morphology*, ed. M. Hildebrand, D.M. Bramble, K.F. Liem, and D.B. Wake (Cambridge, MA: Belknap Press, 1985), 110–28.

8. Humans do create lift with their feet when doing the butterfly stroke.

9. F.E. Fish, "Kinematics and Estimated Thrust Production of Swimming Harp and Ringed Seals," *Journal of Experimental Biology* 137 (1988): 157–73.

10. A.W. English, "Limb Movements and Locomotor Function in the California Sea Lion," *Journal of Zoology, London*, 178 (1976): 341–64. F.E. Fish, "Influence of Hydrodynamic Design and Propulsive Mode on Mammalian Swimming Energetics," *Australian Journal of Zoology* 42 (1993): 79–101.

11. G.C. Hickman, "Swimming Ability in Talpid Moles, with Particular Reference to the Semi-Aquatic Mole *Condylura cristata*," *Mammalia* 48 (1984): 505–13.

12. F.E. Fish, "Transitions from Drag-Based to Lift-Based Propulsion in Mammalian Swimming," *American Zoologist* 36 (1996): 628–41.

13. T.M. Williams, "Locomotion in the North American Mink, a Semi-Aquatic Mammal, I: Swimming Energetics and Body Drag," *Journal of Experimental Zoology* 103 (1983): 155–68.

14. F.E. Fish, "Association of Propulsive Mode with Behavior in River Otters (*Lutra canadensis*)," *Journal of Mammalogy* 75 (1994): 989–97.

15. J.G.M. Thewissen and F.E. Fish, "Locomotor Evolution in the Earliest Cetaceans: Functional Model, Modern Analogues, and Paleontological Evidence," *Paleobiology* 23 (1997): 482–490.

16. S. Bajpai and J.G.M. Thewissen, "A New, Diminuitive Whale from Kachchh (Gujarat, India) and Its Implications for Locomotor Evolution of Cetaceans," *Current Science (New Delhi)* 79 (2000): 1478–82. J.G.M. Thewissen

and S. Bajpai, "New Skeletal Material for *Andrewsiphius* and *Kutchicetus,* Two Eocene Cetaceans from India," *Journal of Paleontology* 83 (2009): 635–63.

17. P.D. Gingerich, "Land-to-Sea Transition in Early Whales: Evolution of Eocene Archaeoceti (Cetacea) in Relation to Skeletal Proportions and Locomotion of Living Semiaquatic Mammals," *Paleobiology* 29 (2003): 429–54.

18. E.A. Buchholtz, "Implications of Vertebral Morphology for Locomotor Evolution in Early Cetacea," in *The Emergence of Whales: Evolutionary Patterns in the Origin of Cetacea,* ed. J.G.M. Thewissen (New York, NY: Plenum Press, 1998), 325–52.

19. R. Dehm and T. zu Oettingen-Spielberg, "Palaeontologische und geologische Untersuchungen im Tertiaer von Pakistan, 2: Die mitteleozaenen Sauegetiere von Ganda Kas bei Basal in Nord-West Pakistan," *Abhandlungen der Bayerischen Akademie der Wissenschaften, Mathematisch.-Naturwissenschaftliche Klasse* 91 (1958): 1–53.

20. S. Bajpai and P.D. Gingerich, "A New Archaeocete (Mammalia, Cetacea) from India and the Time of Origin of Whales," *Proceedings of the National Academy of Sciences* 95 (1998): 15464–68.

21. J.G.M. Thewissen, E.M. Williams, and S.T. Hussain, "Eocene Mammal Faunas from Northern Indo-Pakistan," *Journal of Vertebrate Paleontology* 21 (2001): 347–66.

22. K.K. Smith, "The Evolution of the Mammalian Pharynx," *Zoological Journal of the Linnean Society* 104 (1992): 313–49.

23. J.S. Reidenberg and J.T. Laitman, "Anatomy of the Hyoid Apparatus in Odontoceti (Toothed Whales): Specializations of their Skeleton and Musculature Compared with Those of Terrestrial Mammals," *Anatomical Record* 240 (1994): 598–624.

24. The hyoid of humans is a single bone, located in the midline of the neck, but embryologically, it consists of three bones. In most mammals, there are even more: a dog has nine, for instance.

25. E.J. Slijper, *Whales* (New York, NY: Basic Books, 1962).

26. S. Nummela, S.T. Hussain, and J.G.M. Thewissen, "Cranial Anatomy of Pakicetidae (Cetacea, Mammalia)," *Journal of Vertebrate Paleontology* 26 (2006): 746–59.

27. B. Møhl, W.W.L. Au, J. Pawloski, and P.E. Nachtigall, 1999, "Dolphin Hearing: Relative Sensitivity as a Function of Point of Application of a Contact Sound Source in the Jaw and Head Region," *Journal of the Acoustical Society of America* 105 (1999): 3421–24.

28. S. Nummela, J.G.M. Thewissen, S. Bajpai, T. Hussain, and K. Kumar, "Sound Transmission in Archaic and Modern Whales: Anatomical Adaptations for Underwater Hearing," *Anatomical Record* 290 (2007):716–33. S. Nummela, J.G.M. Thewissen, S. Bajpai, S.T. Hussain, and K.K. Kumar, "Eocene Evolution of Whale Hearing," *Nature* 430 (2004): 776–78.

29. S.I. Madar, J.G.M. Thewissen, and S.T. Hussain, "Additional Holotype Remains of *Ambulocetus natans* (Cetacea, Ambulocetidae), and Their Implications for Locomotion in Early Whales," *Journal of Vertebrate Paleontology* 22 (2002): 405–22.

30. Y. Narita and S. Kuratani, "Evolution of the Vertebral Formulae in Mammals: A Perspective on Developmental Constraints," *Journal of Experimental Zoology Part B: Molecular and Developmental Evolution* 15 (2005): 91–106. J. Müller, T.M. Scheyer, J.J. Head, P.M. Barrett, I. Werneburg, P.G. Ericson, D. Polly, and M.R. Sánchez-Villagra, "Homeotic Effects, Somitogenesis and the Evolution of Vertebral Numbers in Recent and Fossil Amniotes," *Proceedings of the National Academy of Sciences* 107 (2010): 2118–23.

31. M.M. Moran, S. Bajpai, J. C. George, R. Suydam, S. Usip, and J. G. M. Thewissen, "Intervertebral and Epiphyseal Fusion in the Postnatal Ontogeny of Cetaceans and Terrestrial Mammals," *Journal of Mammalian Evolution* (2014), doi:10.1007/s10914-014-9256-7.

32. J.G.M. Thewissen, S.I. Madar, and S.T. Hussain, "*Ambulocetus natans,* an Eocene Cetacean (Mammalia) from Pakistan," *Courier Forschungs.-Institut Senckenberg* 190 (1996): 1–86. L.J. Roe, J.G.M. Thewissen, J. Quade, J.R. O'Neil, S. Bajpai, A. Sahni, and S.T. Hussain, "Isotopic Approaches to Understanding the Terrestrial to Marine Transition of the Earliest Cetaceans," in *The Emergence of Whales: Evolutionary Patterns in the Origin of Cetacea,* ed. J.G.M. Thewissen (New York, NY: Plenum Press, 1998), 399–421. S.I. Madar, J.G.M. Thewissen, and S.T. Hussain, "Additional Holotype Remains of *Ambulocetus natans* (Cetacea, Ambulocetidae), and their Implications for Locomotion in Early Whales," *Journal of Vertebrate Paleontology* 22 (2002): 405–22.

33. D. Gish, "When Is a Whale a Whale?" *Acts & Facts* 23 (1994, No. 4). http://www.icr.org/article/when-whale-whale/.

34. K. Miller, *Finding Darwin's God: A Scientist's Search for Common Ground between God and Evolution* (New York, NY: HarperCollins, 1999).

35. L. Van Valen, "Deltatheridia: A New Order of Mammals," *Bulletin of the American Museum of Natural History* 132 (1966): 1–126.

36. M. Goodman, J. Czelusniak, and J.E. Beeber, "Phylogeny of the Primates and Other Eutherian Orders: A Cladistics Analysis Using Amino Acids and Nucleotide Sequence Data," *Cladistics* 1 (1985): 171–85.

CHAPTER 5. WHEN THE MOUNTAINS GREW

1. The term *Himalayas* is used in two different senses. It refers loosely to all the mountains on the northern side of India, Pakistan, and Bangladesh. More specifically, it refers to one particular mountain range in that area, with a geological history that is very different from the others.

2. University of California Museum of Paleontology, "Alfred Wegener (1880–1930)," http://www.ucmp.berkeley.edu/history/wegener.html.

3. G.E. Pilgrim, "Middle Eocene Mammals from Northwest India," *Proceedings of the Zoological Society* 110 (1940): 124–52.

4. R. Dehm and T. zu Oettingen-Spielberg, "Paläontologische und geologische Untersuchungen im Tertiär von Pakistan, 2: Die mitteleozänen Säugetiere von Ganda Kas bei Basal in Northwest Pakistan," *Abhandlungen der Bayerischen Akademie der Wissenschaften, Mathematisch.-Naturwissenschaftliche Klasse* 91 (1958): 1–54.

5. R. M. West, "Middle Eocene Large Mammal Assemblage with Tethyan Affinities, Ganda Kas Region, Pakistan," *Journal of Paleontology* 54 (1980): 508–33.

6. J.G.M. Thewissen, S.I. Madar, and S.T. Hussain, 1996, "*Ambulocetus natans,* an Eocene Cetacean (Mammalia) from Pakistan," *Courier Forschungs-Institut Senckenberg* 190 (1996): 1–86. Some years later we were able to go back and to excavate the remainder of the holotype of *Ambulocetus natans.* Those fossils are described in S.I. Madar, J.G.M. Thewissen, and S.T. Hussain, "Additional Holotype Remains of *Ambulocetus natans* (Cetacea, Ambulocetidae), and Their Implications for Locomotion in Early Whales," *Journal of Vertebrate Paleontology* 22 (2002): 405–22.

7. A. Sahni, "Enamel Ultrastructure of Fossil Mammalia: Eocene Archaeoceti from Kutch," *Journal of the Palaeontological Society of India* 25 (1981): 33–37.

8. M.C. Maas and J.G.M. Thewissen, "Enamel Microstructure of *Pakicetus* (Mammalia: Archaeoceti)," *Journal of Paleontology* 69 (1995): 1154–63.

CHAPTER 6. PASSAGE TO INDIA

1. Panjab is a state in India; Punjab is a province of Pakistan. When the British ruled India, these were one; when the country broke into two, the province was divided, too.

2. A.B. Wynne, "Memoir on the Geology of Kutch," *Memoirs of the Geological Survey of India* 9 (1872).

3. A. Sahni and V.P. Mishra, "A New Species of *Protocetus* from the Middle Eocene of Kutch, Western India," *Palaeontology* 15 (1972): 490–95.

4. A. Sahni and V.P. Mishra, "Lower Tertiary Vertebrates from Western India," *Monographs of the Palaeontological Society of India* 3 (1975).

5. R. Kellogg, *A Review of the Archaeoceti* (Washington, DC: Carnegie Institute of Washington, 1936).

6. S. Bajpai and J.G.M. Thewissen, "Middle Eocene Cetaceans from the Harudi and Subathu Formations of India," in *The Emergence of Whales: Evolutionary Patterns in the Origin of Cetacea,* ed. J.G.M. Thewissen (New York, NY: Plenum Press, 1998), 213–34.

CHAPTER 7. A TRIP TO THE BEACH

1. S.K. Biswas, "Tertiary Stratigraphy of Kutch," *Memoirs of the Geological Society of India* 10 (1992): 1–29.

2. S.K. Mukhopadhyay and S. Shome, "Depositional Environment and Basin Development during Early Paleaeogene Lignite Deposition, Western Kutch, Gujarat," *Journal of the Geological Society of India* 47 (1996): 579–92.

CHAPTER 8. THE OTTER WHALE

1. S. Bajpai and J. G. M. Thewissen, "A New, Diminuitive Whale from Kachchh (Gujarat, India) and Its Implications for Locomotor Evolution of Cetaceans," *Current Science (New Delhi)* 79 (2000): 1478–82.

2. A. Sahni and V.P. Mishra, "Lower Tertiary Vertebrates from Western India," *Monographs of the Palaeontological Society of India* 3 (1975).

3. K. Kumar and A. Sahni, "*Remingtonocetus harudiensis*: New Combination, a Middle Eocene Archaeocete (Mammalia, Cetacea) from Western Kutch, India," *Journal of Vertebrate Paleontology* 6 (1986): 326–49.

4. P.D. Gingerich, M. Arif, and W.C. Clyde, "New Archaeocetes (Mammalia, Cetacea) from the Middle Eocene Domanda Formation of the Sulaiman Range, Punjab, Pakistan," *Contributions of the Museum of Paleontology, University of Michigan* 29 (1995): 291–330.

5. J.G.M. Thewissen and S. Bajpai, "Dental Morphology of the Remingtonocetidae (Cetacea, Mammalia)," *Journal of Paleontology* 75 (2001): 463–65.

6. J.G.M. Thewissen and S. T. Hussain, "*Attockicetus praecursor,* a New Remingtonocetid Cetacean from Marine Eocene Sediments of Pakistan," *Journal of Mammalian Evolution* 7 (2000): 133–46.

7. V. Ravikant and S. Bajpai, "Strontium Isotope Evidence for the Age of Eocene Fossil Whales of Kutch, Western India," *Geological Magazine* 147 (2012): 473–77.

8. P.D. Gingerich, M. Ul-Haq, W. V. Koenigswald, W.J. Sanders, B.H. Smith, and I.S. Zalmout, "New Protocetid Whale from the Middle Eocene of Pakistan: Birth on Land, Precicial Development, and Sexual Dimorphism," *PLoS One* 4 (2009): E4366.

9. L.N. Cooper, T.L. Hieronymus, C.J. Vinyard, S. Bajpai, and J.G.M. Thewissen, "Feeding Strategy in Remingtonocetinae (Cetacea, Mammalia) by Constrained Ordination," in *Experimental Approaches to Understanding Fossil Organisms*, ed. D. I. Hembree, B. F. Platt, and J. J. Smith (Dordrecht, Plenum, 2014), 89–107.

10. If the teeth had fallen out during life, the space in the jaw (the alveolus) that the tooth was anchored in would be filled by new bone.

11. J.G.M. Thewissen and S. Bajpai, "New Skeletal Material of *Andrewsiphius* and *Kutchicetus,* Two Eocene Cetaceans from India," *Journal of Paleontology* 83 (2009): 635–63.

12. R. Elsner, "Living in Water: Solutions to Physiological Problems," in *Biology of Marine Mammals*, ed. J.E. Reynolds III and S.A. Rommel (Washington, DC: Smithsonian Institution Press), 73–116.

13. S. Nummela, S.T. Hussain, and J.G.M. Thewissen, "Cranial Anatomy of Pakicetidae (Cetacea, Mammalia)," *Journal of Vertebrate Paleontology* 26 (2006): 746–59.

14. R.M. Bebej, M. Ul-Haq, I.S. Zalmout, and P.D. Gingerich, "Morphology and Function of the Vertebral Column in *Remingtonocetus domandaensis* (Mammalia, Cetacea) from the Middle Eocene Domanda Formation of Pakistan," *Journal of Mammalian Evolution* 19 (2012): 77–104. doi:10.1007 /S10914-011-9184-8.

15. F. Spoor, S. Bajpai, S.T. Hussain, K. Kumar, and J.G.M. Thewissen, "Vestibular Evidence for the Evolution of Aquatic Behaviour in Early Cetaceans," *Nature* 417 (2002): 163–66.

16. A. Williams and J. Safarti, "Not at All Like a Whale," *Creation* 27 (2005): 20–22.

CHAPTER 9. THE OCEAN IS A DESERT

1. K. Schmidt-Nielsen, *Animal Physiology: Adaptation and Environment* (Cambridge: Cambridge University Press, 1997).

2. M.E.Q. Pilson, "Water Balance in California Sea Lions," *Physiological Zoology* 43 (1970): 257–69.

3. D.P. Costa, "Energy, Nitrogen, Electrolyte Flux and Sea Water Drinking in the Sea Otter *Enhydra lutris*," *Physiological Zoology* 55 (1982): 35–44.

4. R.M. Ortiz, "Osmoregulation in Marine Mammals," *Journal of Experimental Biology* 204 (2001): 1831–44.

5. C. Hui, "Seawater Consumption and Water Flux in the Common Dolphin *Delphinus delphis*," *Physiological Zoology* 54 (1981): 430–40.

6. J.G.M. Thewissen, L.J. Roe, J.R. O'Neil, S.T. Hussain, A. Sahni, and S. Bajpai, "Evolution of Cetacean Osmoregulation," *Nature* 381 (1996): 379–80.

7. M.T. Clementz, A. Goswami, P.D. Gingerich, and P.L. Koch, "Isotopic Records from Early Whales and Seacows: Contrasting Patterns of Ecological Transition," *Journal of Vertebrate Paleontology* 26 (2006): 355–70.

8. L.J. Roe, J.G.M. Thewissen, J. Quade, J.R. O'Neil, S. Bajpai, A. Sahni, and S.T. Hussain, "Isotopic Approaches to Understanding the Terrestrial-to-Marine Transition of the Earliest Cetaceans," in *The Emergence of Whales: Evolutionary Patterns in the Origin of Cetacea*, ed. J.G.M. Thewissen (New York, NY: Plenum, 1998), 399–422.

CHAPTER 10. THE SKELETON PUZZLE

1. L. Van Valen, "Deltatheridia, a New Order of Mammals," *Bulletin of the American Museum of Natural History* 132 (1966): 1–126.

2. X. Zhou, R. Zhai, P. Gingerich, and L. Chen, "Skull of a New Mesonychid (Mammalia, Mesonychia) from the Late Paleocene of China," *Journal of Vertebrate Paleontology* 15 (2009): 387–400.

3. Z. Luo and P.D. Gingerich, "Terrestrial Mesonychia to Aquatic Cetacea: Transformation of the Basicranium and Evolution of Hearing in Whales," *University of Michigan Papers on Paleontology* 31 (1999), 1–98. M.A. O'Leary and J.H. Geisler, "The Position of Cetacea within Mammalia: Phylogenetic Analysis of Morphological Data from Extinct and Extant Taxa," *Systematic Biology* 48 (1999): 455–90. M.D. Uhen, "New Species of Protocetid Archaeocete Whale, *Eocetus wardii* (Mammalia, Cetacea) from the Middle Eocene of North Carolina," *Journal of Paleontology* 73 (1999): 512–28.

4. J.G.M. Thewissen, E.M. Williams, L.J. Roe, and S.T. Hussain, "Skeletons of Terrestrial Cetaceans and the Relationship of Whales to Artiodactyls," *Nature* 413 (2001): 277–81.

5. See note 4.

6. P.D. Gingerich, M.U. Haq, I.S. Zalmout, I.H. Khan, and M.S. Malkani, "Origin of Whales from Early Artiodactyls: Hands and Feet of Eocene Protocetidae from Pakistan," *Science* 293 (2001): 2239–42.

7. M.C. Milinkovitch, M. Bérubé, and P.J. Palsbøl, "Cetaceans Are Highly Derived Artiodactyls," in *The Emergence of Whales: Evolutionary Patterns in*

the Origin of Cetacea, ed. J.G.M. Thewissen (New York, NY: Plenum Press, 1998), 113–131. M. Nikaido, A.P. Rooney, and N. Okada, "Phylogenetic Relationships among Cetartiodactyls Based on Insertions of Short and Long Interspersed Elements: Hippopotamuses Are the Closest Extant Relatives of Whales," *Proceedings of the National Academy of Sciences* 96 (1999): 10261–66. J. Gatesy and M.A. O'Leary, "Deciphering Whale Origins with Molecules and Fossils," *Trends in Ecology and Evolution* 16 (2001): 562–70.

CHAPTER 11. THE RIVER WHALES

1. K.S. Norris, "The Evolution of Acoustic Mechanisms in Odontocete Cetaceans," in *Evolution and Environment*, ed. E.T. Drake (New Haven, CT: Yale University Press, 1968), 297–324. T.W. Cranford, P. Krysl, and J.A. Hildebrand, "Acoustic Pathways Revealed: Simulated Sound Transmission and Reception in Cuvier's Beaked Whale (*Ziphius cavirostris*)," *Bioinspiration and Biomimetics* 3 (2008): 016001. doi:10.1088/1748-3182/3/1/016001.

2. J.G. McCormick, E.G. Wever, G. Palin, and S.H. Ridgway, "Sound Conduction in the Dolphin Ear," *Journal of the Acoustical Society of America* 48 (1970): 1418–28.

3. S. Hemilä, S. Nummela, and T. Reuter, "A Model of the Odontocete Middle Ear," *Hearing Research* 133 (1999): 82–97.

4. T.W. Cranford, P. Krysl, and M. Amundin, "A New Acoustic Portal into the Odontocete Ear and Vibrational Analysis of the Tympanoperiotic Complex," *PLoS One* 5 (2010): E11927. doi:10.1371/Journal.Pone.0011927.

5. W.C. Lancaster, "The Middle Ear of the Archaeoceti," *Journal of Vertebrate Paleontology* 10 (1990): 117–27. S. Nummela, J.G.M. Thewissen, S. Bajpai, S.T. Hussain, and K. Kumar, "Eocene Evolution of Whale Hearing," *Nature* 430 (2004): 776–78. S. Nummela, J.E. Kosove, T.E. Lancaster, and J.G.M. Thewissen, "Lateral Mandibular Wall Thickness in *Tursiops truncatus*: Variation Due to Sex and Age," *Marine Mammal Science* 20 (2004): 491–97. S. Nummela, J.G.M. Thewissen, S. Bajpai, S.T. Hussain, and K. Kumar, "Sound Transmission in Archaic and Modern Whales: Anatomical Adaptations for Underwater Hearing," *Anatomical Record* 290 (2007): 716–33.

6. D.M. Higgs, E.F. Brittan-Powell, D. Soares, M.J. Souza, C.E. Carr, R.J. Dooling, and A.N. Popper, "Amphibious Auditory Responses of the American Alligator *(Alligator mississippiensis)*," *Journal of Comparative Physiology* 188 (2002): 217–23.

7. R. Rado, M. Himelfarb, B. Arensburg, J. Terkel, and Z. Wollberg, "Are Seismic Communication Signals Transmitted by Bone Conduction in the Blind Mole Rat?" *Hearing Research* 41 (1989): 23–29.

8. Modern whales also can still hear in air—in spite of not having a functional eardrum or external auditory meatus—but their underwater hearing is much better.

9. S. Nummela, J.E. Kosove, T. Lancaster, and J.G.M. Thewissen. "Lateral Mandibular Wall Thickness in *Tursiops truncatus*: Variation Due to Sex and Age," *Marine Mammal Science* 20 (2004): 491–97.

10. R.M. West, "Middle Eocene Large Mammal Assemblage with Tethyan Affinities, Ganda Kas Region, Pakistan," *Journal of Paleontology* 54 (1980): 508–33. P.D. Gingerich and D.E. Russell, "*Pakicetus inachus,* a New Archaeocete (Mammalia, Cetacea)," *Contributions from the Museum of Paleontology, University of Michigan* 25 (1981): 235–46. K. Kumar and A. Sahni, "Eocene Mammals from the Upper Subathu Group, Kashmir Himalaya, India," *Journal of Vertebrate Paleontology* 5 (1985): 153–68. J.G.M. Thewissen and S.T. Hussain, "Systematic Review of the Pakicetidae, Early and Middle Eocene Cetacea (Mammalia) from Pakistan and India," *Bulletin of the Carnegie Museum of Natural History* 34 (1998): 220–38.

11. S. I. Madar, "The Postcranial Skeleton of Early Eocene Pakicetid Cetaceans," *Journal of Paleontology* 81 (2007): 176–200.

12. L.J. Roe, J.G.M. Thewissen, J. Quade, J.R. O'Neil, S. Bajpai, A. Sahni, and S.T. Hussain, "Isotopic Approaches to Understanding the Terrestrial to Marine Transition of the Earliest Cetaceans," in *The Emergence of Whales: Evolutionary Patterns in the Origin of Cetacea,* ed. J.G.M. Thewissen (New York, NY: Plenum Press, 1998), 399–421. M.T. Clementz, A. Goswami, P.D. Gingerich, and P.L. Koch, "Isotopic Records from Early Whales and Sea Cows: Contrasting Patterns of Ecological Transition," *Journal of Vertrebrate Paleontology* 26 (2006): 355–70.

13. M.A. O'Leary and M.D. Uhen, "The Time of Origin of Whales and the Role of Behavioral Changes in the Terrestrial–Aquatic Transition," *Paleobiology* 25 (1999): 534–56. J.G.M. Thewissen, M.T. Clementz, J.D. Sensor, and S. Bajpai, "Evolution of Dental Wear and Diet During the Origin of Whales," *Paleobiology* 37 (2011): 655–69.

14. P.S. Ungar, *Mammal Teeth: Origin, Evolution, and Diversity* (Baltimore, MD: Johns Hopkins Press, 2010).

15. A. D. Foote, J. Newton, S.B. Piertney, E. Willerslev, and M.T.P. Gilbert, "Ecological, Morphological, and Genetic Divergence of Sympatric North Atlantic Killer Whale Populations," *Molecular Ecology* 18 (2009): 5207–17.

16. S. Nummela, S.T. Hussain, and J.G.M. Thewissen, "Cranial Anatomy of Pakicetidae (Cetacea, Mammalia)," *Journal of Vertebrate Paleontology* 26 (2006), 746–59.

17. G. Dehnhardt and B. Mauck, "Mechanoreception in Secondarily Aquatic Vertebrates," in *Sensory Evolution on the Threshold: Adaptations in Secondarily Aquatic Vertebrates,* ed. J.G.M. Thewissen and S. Nummela (Berkeley, CA: University of California Press, 2008), 295–316.

18. N.M. Gray, K. Kainec, S. Madar, L. Tomko, and S. Wolfe, "Sink or Swim? Bone Density As a Mechanism for Buoyancy Control in Early Cetaceans," *Anatomical Record* 290 (2007): 638–53.

19. S.I. Madar, "The Postcranial Skeleton of Early Eocene Pakicetid Cetaceans," *Journal of Vertebrate Paleontology* 81 (2007): 176–200.

20. See note 12.

21. J. G. M. Thewissen, L. N. Cooper, M. T. Clementz, S. Bajpai, and B. N. Tiwari. "Whales Originated from Aquatic Artiodactyls in the Eocene Epoch of India," *Nature* 450 (2007): 1190–95.

CHAPTER 12. WHALES CONQUER THE WORLD

1. M. Nikaido, A. P. Rooney, and N. Okada, "Phylogenetic Relationships among Cetartiodactyls Based on Insertions of Short and Long Interspersed Elements: Hippopotamuses Are the Closest Extant Relatives of Whales," *Proceedings of the National Academy of Sciences* 96 (1999): 10261–66.

2. J.-R. Boisserie, F. Lihoreau, and M. Brunet, "The Position of Hippopotamidae within Cetartiodactyla," *Proceedings of the National Academy of Sciences* 102 (2005): 1537–41.

3. J. G. M. Thewissen and S. Bajpai, "New Protocetid Cetaceans from the Eocene of India," *Palaeontologia Electronica* (in review).

4. Paleobiology Database, http://fossilworks.org/?a=home.

5. A. Sahni and V. P. Mishra, "Lower Tertiary Vertebrates from Western India," *Monograph of the Palaeontological Society of India* 3 (1975): 1–48. P. D. Gingerich, M. Arif, M. A. Bhatti, M. Anwar, and W. J. Sanders, "*Protosiren* and *Babiacetus* (Mammalia, Sirenia and Cetacea) from the Middle Eocene Drazinda Formation, Sulaiman Range, Punjab (Pakistan)," *Contributions from the Museum of Paleontology, University of Michigan* 29 (1995): 331–57. P. D. Gingerich, M. Arif, and W. C. Clyde, "New Archaeocetes (Mammalia, Cetacea) from the Middle Eocene Domanda Formation of the Sulaiman Range, Punjab (Pakistan)," *Contributions from the Museum of Paleontology, University of Michigan* 29 (1995): 291–330. P. D. Gingerich, M. Haq, I. S. Zalmout, I. H. Khan, and M. S. Malkani, "Origin of Whales from Early Artiodactyls: Hands and Feet of Eocene Protocetidae from Pakistan," *Science* 293 (2001): 2239–42. P. D. Gingerich, M. ul-Haq, W. v. Koenigswald, W. J. Sanders, B. H. Smith, and I. S. Zalmout, "New Protocetid Whale from the Middle Eocene of Pakistan: Birth on Land, Precocial Development, and Sexual Dimorphism," *PLoS One* 4 (2009): e4366, doi:10.1371/journal.pone.0004366.

6. E. M. Williams, "Synopsis of the Earliest Cetaceans: Pakicetidae, Ambulocetidae, Remingtonocetidae, and Protocetidae," in *Emergence of Whales: Evolutionary Patterns in the Origin of Cetacea*, ed. J. G. M. Thewissen (New York: Plenum Press, 1988), 1–28. G. Bianucci and P. D. Gingerich, "*Aegyptocetus tarfa* n. gen. et sp. (Mammalia, Cetacea), from the Middle Eocene of Egypt: Clinorhynchy, Olfaction, and Hearing in a Protocetid Whale," *Journal of Vertebrate Paleontology* 31 (2011): 1173–88. P. D. Gingerich, "Cetacea," in *Cenozoic Mammals of Africa*, ed. L. Werdelin and W. J. Sanders (Berkeley: University of California Press, 2010), 873–99.

7. R. C. Hulbert, Jr., R. M. Petkewich, G. A. Bishop, D. Bukry, and D. P. Aleshire, "A New Middle Eocene Protocetid Whale (Mammalia: Cetacea: Archaeoceti) and Associated Biota from Georgia," *Journal of Paleontology* 72 (1998): 907–26. J. H. Geisler, A. E. Sanders, and Z.-X. Luo, "A New Protocetid Whale (Cetacea: Archaeoceti) from the Late Middle Eocene of South Carolina," *American Museum Novitates* 3480 (2005): 1–65. S. A. McLeod and L. G. Barnes, "A New Genus and Species of Eocene Protocetid Archaeocete Whale (Mammalia, Cetacea) from the Atlantic Coastal Plain," *Science Series, Natural History Museum of Los Angeles County* 41 (2008): 73–98. M. D. Uhen, "New Specimens of Protocetidae (Mammalia, Cetacea) from New Jersey and South Carolina," *Journal of Vertebrate Paleontology* 34 (2013): 211–19.

8. M. D. Uhen, N. D. Pyenson, T. J. Devries, M. Urbina, and P. R. Renne, "New Middle Eocene Whales from the Pisco Basin of Peru," *Journal of Paleontology* 85 (2011): 955–69.

9. M. T. Clementz, A. Goswami, P. D. Gingerich, and P. L. Koch, "Isotopic Records from Early Whales and Sea Cows: Contrasting Patterns of Ecological Transition," *Journal of Vertebrate Paleontology* 26 (2006): 355–70.

10. See figure 50.

11. S. Bajpai and J. G. M. Thewissen, 1998, "Middle Eocene Cetaceans from the Harudi and Subathu Formations of India," in *Emergence of Whales: Evolutionary Patterns in the Origin of Cetacea*, ed. J. G. M. Thewissen (New York: Plenum Press, 1988), 213–33.

12. P. D. Gingerich, M. ul-Haq, I. H. Khan, and I. S. Zalmout, "Eocene Stratigraphy and Archaeocete Whales (Mammalia, Cetacea) of Drug Lahar in the Eastern Sulaiman Range, Balochistan (Pakistan)," *Contributions from the Museum of Paleontology, University of Michigan* 30 (2001): 269–319.

13. P. D. Gingerich, I. S. Zalmout, M. ul-Haq, and M. A. Bhatti, "*Makaracetus bidens*, a New Protocetid Archaeocete (Mammalia, Cetacea) from the Early Middle Eocene of Balochistan (Pakistan)," *Contributions from the Museum of Paleontology, University of Michigan* 31 (2005): 197–210.

14. J. G. M. Thewissen, J. C. George, C. Rosa, and T. Kishida, "Olfaction and Brain Size in the Bowhead Whale (*Balaena mysticetus*)," *Marine Mammal Science* 27 (2011): 282–94.

15. H. H. A. Oelschläger and J. S. Oelschläger, "Brain," in *Encyclopedia of Marine Mammals* (1st ed.), ed. W. F. Perrin, B. Würsig, and J. G. M. Thewissen (San Diego, CA: Academic Press, 2002), 133–58.

16. S. J. Godfrey, J. Geisler, and E. M. G. Fitzgerald, "On the Olfactory Anatomy in an Archaic Whale (Protocetidae, Cetacea) and the Minke Whale *Balaenoptera acutorostrata* (Balaenopteridae, Cetacea)," *Anatomical Record* 296 (2013): 257–72.

17. T. Edinger, "Hearing and Smell in Cetacean History," *Monatschrift für Psychiatrie und Neurologie* 129 (1955): 37–58.

18. P. A. Brennan and F. Zufall, "Pheromonal Communication in Vertebrates," *Nature* 444 (2006): 308–15.

19. J. Henderson, R. Altieri, and D. Müller-Schwarze, "The Annual Cycle of Flehmen in Black-Tailed Deer (*Odocoileus hemionis columbianus*)," *Journal of Chemical Ecology* 6 (1980): 537–57.

20. J. E. King, *Seals of the World* (New York: Cornell University Press, 1983).

21. R. A. Dart, "The Brain of the Zeuglodontidae (Cetacea)," *Proceedings of the Zoological Society, London* 42 (1923): 615–54.

22. S. Bajpai, J. G. M. Thewissen, and A. Sahni. "*Indocetus* (Cetacea, Mammalia) Endocasts from Kachchh (India)," *Journal of Vertebrate Paleontology* 16 (1996): 582–84

23. H. J. Jerison, *Evolution of the Brain and Intelligence* (New York: Academic Press, 1973).

24. L. Marino, "Cetacean Brain Evolution: Multiplication Generates Complexity," *International Journal of Comparative Psychology* 17 (2004): 1–16.

25. In sea lions, which swim with their forelimbs, the longest finger is 1.7 times as long as the first part of the limb (the humerus). In seals, which swim with their hind limb, the longest toe/femur ratio is 2.4. In *Ambulocetus*, this ratio is 1.1, and in the protocetids *Rodhocetus* and *Maiacetus* it is 0.95 and 0.79, respectively. The small foot of the protocetids suggests that it is less involved in propulsion than that of ambulocetids.

26. A. W. English. "Limb Movements and Locomotor Function in the California Sea Lion (*Zalophus californianus*)," *Journal of the Zoological Society of London* 178 (1976): 341–64.

27. P. D. Gingerich, "Land-to-Sea Transition of Early Whales: Evolution of Eocene Archaeoceti (Cetacea) in Relation to Skeletal Proportions and Locomotion of Living Semiaquatic Mammals," *Paleobiology* 29 (2003): 429–54.

28. M. D. Uhen, "Form, Function, and Anatomy of *Dorudon atrox* (Mammalia, Cetacea): An Archaeocete from the Middle to Late Eocene of Egypt," *University of Michigan, Papers on Paleontology* 34 (2004): 1–222.

29. V. de Buffrénil, A. de Ricqlès, C. E. Ray, and D. P. Domning, "Bone Histology of the Ribs of the Archaeocetes (Mammalia, Cetacea)," *Journal of Vertebrate Paleontology* 10 (1990): 455–66.

30. Y. Narita and S. Kuratani, "Evolution of the Vertebral Formulae in Mammals: A Perspective on Developmental Constraints," *Journal of Experimental Zoology B: Molecular and Developmental Evolution* 15 (2005): 91–106.

31. J. G. M. Thewissen, L. N. Cooper, and R. R. Behringer, "Developmental Biology Enriches Paleontology," *Journal of Vertebrate Paleontology* 32 (2012): 1224–34.

32. E. M. Williams, "Synopsis of the Earliest Cetaceans: Pakicetidae, Ambulocetidae, Remingtonocetidae, and Protocetidae," in *Emergence of Whales: Evolutionary Patterns in the Origin of Cetacea*, ed. J. G. M. Thewissen (New York: Plenum Press, 1988), 1–28.

33. P. D. Gingerich, M. ul-Haq, W. v. Koenigswald, W. J. Sanders, B. H. Smith, and I. S. Zalmout, "New Protocetid Whale from the Middle Eocene of Pakistan: Birth on Land, Precocial Development, and Sexual Dimorphism, *PLoS One* 4 (2009): e4366, doi:10.1371/journal.pone.0004366.

34. E. Fraas, "Neue Zeuglodonten aus dem unteren Mitteleozän von Mokattam bei Cairo," *Geologische und Paläontologische Abhandlungen* 6 (1904): 199–220.

CHAPTER 13. FROM EMBRYOS TO EVOLUTION

1. This hunt was exposed, years later, in the Academy Award–winning movie *The Cove*.

2. Reviewed in L. Bejder and B.K. Hall, "Limbs in Whales and Limblessness in Other Vertebrates: Mechanisms of Evolutionary and Developmental Transformation and Loss," *Evolution and Development* 4 (2002): 445–58.

3. R.C. Andrews, "A Remarkable Case of External Hind Limbs in a Humpback Whale," *American Museum Novitates* 9 (1921): 1–6.

4. At the time of writing, Haruka was alive, but the dolphin died in April, 2013.

5. R. O'Rahilly and F. Müller, *Developmental Stages in Human Embryos* (Washington, DC: Carnegie Institute of Washington, 1987).

6. Proteins often have remarkably inappropriate, cumbersome, or silly names, so in publications they are usually just referred to by a letter–number combination such as this one.

7. J.-D. Bénazet and R. Zeller, "Vertebrate Limb Development: Moving from Classical Morphogen Gradients to an Integrated 4-dimensional Patterning System," *Cold Spring Harbor Perspectives on Biology* 1(2009): a001339.

8. B.D. Harfe, P.J. Scherz, S. Nissin, H. Tiam, A.P. McMahon, and C.J. Tabin, "Evidence for an Expansion-Based Temporal SHH Gradient in Specifying Vertebrate Digit Identities," *Cell* 118 (2004): 517–28.

9. L.N. Cooper, A. Berta, S.D. Dawson, and J.S. Reidenberg, "Evolution of Hyperphalangy and Digit Reduction in the Cetacean Manus," *Anatomical Record* 290 (2007): 654–72.

10. W. Kükenthal, "Vergleichend anatomische und entwicklungsgeschichtliche Untersuchungen an Waltieren," *Denkschrifte der Medizinische-Naturwissenschaftliche Gesellschaft, Jena* 75 (1893): 1–448.

11. G. Guldberg and F. Nansen, *On the Development and Structure of the Whale, Part 1: On the Development of the Dolphin* (Bergen, Norway: J. Grieg, 1894).

12. W. Kükenthal, "Ueber Rudimente von Hinterflosse bei Embryonen von Walen," *Anatomischer Anzeiger* (1895): 534–37.

13. E. Bresslau, *The Mammary Apparatus of the Mammalia in the Light of Ontogenesis and Phylogenesis* (London: Methuen, 1920).

14. G. Guldberg, "Neue Untersuchungen über die Rudimente von Hinterflossen und die Milchdrüsenanlage bei jungen Delphinenembryonen," *Internationales Monatschrift für Anatomie und Physiologie* 4 (1899): 301–20.

15. M.S. Anderssen, "Studier over mammarorganernes utvikling hos *Phocaena communis*," *Bergens Museum Aarbok, Naturvidensk. R.* 3 (1917–1918): 1–45. http://www.biodiversitylibrary.org/item/130733#page/

16. J.G.M. Thewissen, M.J. Cohn, L.S. Stevens, S. Bajpai, J. Heyning, and W.E. Horton, Jr., "Developmental Basis for Hind-Limb Loss in Dolphins and the Origin of the Cetacean Bodyplan," *Proceedings of the National Academy of Sciences* 103 (2007): 8414–18.

17. M.D. Shapiro, J. Hanken, and N. Rosenthal, "Developmental Basis of Evolutionary Digit Loss in the Australian Lizard *Hemiergis*," *Journal of Experimental Zoology* 297 (2003): 48–57.

18. H. Ito, K. Koizumi, H. Ichishima, S. Uchida, K. Hayashi, K. Ueda, Y. Uezu, , H. Shirouzu, T. Kirihata, M. Yoshioka, S. Ohsumi, and H. Kato, "Inner Structure of the Fin-Shaped Hind Limbs of a Bottlenose Dolphin (*Tursiops truncatus*)," *Abstracts, Biennial Conference on the Biology of Marine Mammals, Tampa, Florida* (2011), 142.

CHAPTER 14. BEFORE WHALES

1. J.H. Geisler and M.D. Uhen, "Morphological Support for a Close Relationship between Hippos and Whales," *Journal of Vertebrate Paleontology* 23 (2003): 991–96.

2. A. Ranga Rao, "New Mammals from Murree (Kalakot Zone) of the Himalayan Foot Hills Near Kalakot, Jammu & Kashmir State, India," *Journal of the Geological Society of India* 12 (1971): 125–34. A. Ranga Rao, "Further Studies on the Vertebrate Fauna of Kalakot, India," *Directorate of Geology, Oil and Natural Gas Commission, Dehradun, Special Paper* 1 (1972): 1–22.

3. J.G.M. Thewissen, L.N. Cooper, M.T. Clementz, S. Bajpai, and B.N. Tiwari, "Whales Originated from Aquatic Artiodactyls in the Eocene Epoch of India," *Nature* 450 (2007): 1190–94.

4. A. Sahni and S.K. Khare, "Three New Eocene Mammals from Rajauri District, Jammu and Kashmir," *Journal of the Paleontological Society of India*, 16 (1971): 41–53. A. Sahni and S.K. Khare, "Additional Eocene Mammals from the Subathu Formation of Jammu and Kashmir," *Journal of the Palaeontological Society of India* 17 (1973): 31–49. J.G.M. Thewissen, E.M. Williams, and S.T. Hussain, "Eocene Mammal Faunas from Northern Indo-Pakistan," *Journal of Vertebrate Paleontology* 21 (2001): 347–66.

5. J.H. Geisler and J.M. Theodor, "Hippopotamus and Whale Phylogeny," *Nature* 458 (2009): 1–4. J. Gatesy, J.H. Geisler, J. Chang, C. Buell, A. Berta, R.W. Meredith, M.S. Springer, and M.R. McGowen, "Phylogenetic Blueprint for a Modern Whale," *Molecular Phylogeny and Evolution* 66 (2013): 479–506.

6. M. Spaulding, M.A. O'Leary, and J. Gatesy, "Relationships of Cetacea (Artiodactyla) among Mammals: Increased Taxon Sampling Alters Interpretations of Key Fossils and Character Evolution," *Plos One* 4 (2009): E7062.

7. J. Gatesy, J.H. Geisler, J. Chang, C. Buell, A. Berta, R.W. Meredith, M.S. Springer, and M R. McGowen (2013) "A phylogenetic blueprint for a modern whale." *Molecular phylogenetics and evolution* 66:479–506.

8. See note 3.

9. L.N. Cooper, J.G.M. Thewissen, S. Bajpai, and B.N. Tiwari, "Postcranial Morphology and Locomotion of the Eocene Raoellid *Indohyus* (Artiodactyla: Mammalia)," *Historical Biology* 24 (2011): 279–310. http://dx.doi.org/10.1080/08912963.2011.624184.

10. G. Dubost, "Un aperçu sur l'écologie du chevrotain africain *Hyemoschus aquaticus* Ogilby, Artiodactyle Tragulide," *Mammalia* 42 (1978): 1–62. E. Meijaard, U. Umilaela, and G. deSilva Wijeyeratne, "Aquatic Escape Behavior in Mouse-Deer Provides Insights into Tragulid Evolution," *Mammalian Biology* 2009: 1–3.

CHAPTER 15. THE WAY FORWARD

1. A.S. Tucker and P. Sharpe, "The Cutting-Edge of Mammalian Development: How the Embryo Makes Teeth," *Nature Reviews, Genetics* 5 (2004): 499–508.

2. J.T. Streelman and R.C. Albertson, "Evolution of Novelty in the Cichlid Dentition," *Journal of Experimental Zoology Part B: Molecular and Developmental Evolution* 306 (2006): 216–26. G.J. Fraser, R.F. Bloomquist, and J.T. Streelman, "A Periodic Pattern Generator for Dental Diversity," *BMC Biology* 6 (2008): 32. doi:10.1186/1741-7007-6-32.

Notes | 231

3. P. M. Munne, S. Felszeghy, M. Jussila, M. Suomalainen, I. Thesleff, and J. Jernvall, "Splitting Placodes: Effects of Bone Morphogenetic Protein and Activin on the Patterning and Identity of Mouse Incisors," *Evolution and Development* 12 (2010): 383–92.

4. B.A. Armfield, Z. Zheng, S. Bajpai, C.J. Vinyard, and J.G.M. Thewissen, "Development and Evolution of the Unique Cetacean Dentition," *PeerJ* 1 (2013): E24. doi:10.7717/peerj.24.

5. See note 4.

6. K. Karlsen, "Development of Tooth Germs and Adjacent Structures in the Whalebone Whale (*Balaenoptera physalus* L.) with a Contribution to the Theories of the Mammalian Tooth Development," *Hvalradets Skrifter Norske Videnskaps-Akademi Olso* 45 (1962): 1–56.

7. M.C.V. Dissel-Scherft and W. Vervoort, "Development of the Teeth in Fetal *Balaenoptera physalus* (L.) (Cetacea, Mystacoceti)," *Proceedings of the Koninklijke Nederlandse Akademie Der Wetenschappen, Serie C* 57 (1954): 196–210.

8. H. Ishikawa and H. Amasaki, "Development and Physiological Degradation of Tooth Buds and Development of Rudiment of Baleen Plate in Southern Minke Whale, *Balaenoptera acutorostrata*," *Journal of Veterinary Medical Science* 57 (1995): 665–70. H. Ishikawa, H. Amasaki, A. Dohguchi, A. Furuya, and K. Suzuki, "Immunohistological Distributions of Fibronectin, Tenascin, Type I, III and IV Collagens, and Laminin during Tooth Development and Degeneration in Fetuses of Minke Whale, *Balaenoptera acutorostrata*," *Journal of Veterinary Medical Science* 61 (1999): 227–32.

9. T.A. Demèrè, M.R. McGowen, A. Berta, and J. Gatesy, "Morphological and Molecular Evidence for a Stepwise Evolutionary Transition from Teeth to Baleen in Mysticete Whales," *Systematic Biology* 57 (2008): 15–37.

Index

Buckley, S.B., 14
Buell, Carl, 108, 109, 116
buffalo, 99, 101, 151
bulla, 4–6, 5f, 47. *See also* tympanic bone

Caenotherium, an artiodactyl, 186f
calcium carbonate, 40–41
camel, 202f
canine, tooth, 21, 21b, 22f, 37, 59, 111,
    147f, 209
Cape Cod, King Lizard of, 9–17, 23
carbon isotopes, 132, 203, 205
Carnegie stages, 177, 180f, 181
Carolinacetus, 163
carpals, 28f
cartilage in limb development, 177
cats, 24b, 113
catfish, 162, 166
cattle, *see* cow
caudal oscillation, 53–54, 55f, 169, 186
caudal vertebra, 26, 27f, 161, 169. *Synonym*
    tail vertebra
caudal undulation, 56; dorsoventral, 55f;
    lateral, 55f
centrum, of vertebra, 13f, 26, 27, 27f.
    Synonym body of vertebra
Cenozoic era, 32f
cerebellum: of protocetids, 168; of
    remingtonocetids, 114, 115f
cerebrum: of protocetids, 168; of reming-
    tonocetids, 114, 115f
cervical vertebrae, *see* neck vertebrae
Cetacea. *See* cetacean
cetacean: artiodactyls as relatives of, 65,
    132–36, 200–203, 201f; classification
    of, 215n17; defined, 5f, 6, 213n6;
    hippos as relatives of, 65, 130, 157, 159,
    200–201; suborders of, 9; transition
    from land to water by, 207–9
Cetacean Research Institute, 188, 189
Chamberlin, Rollin T., 70
Chandigarh, India, 79–80, 82, 85
character matrix, 133–34, 134b
chewing, 61b, 64, 109, 113, 146–48
chimpanzee, 24b
Chocolate Limestone, 87, 88, 88f, 89, 96,
    97–98
Chorgali Formation, 200
cladistic analysis, 133–35, 134–35b
cladograms: of ear ossicle evolution, 7f;
    123f; of habitat and drinking behavior
    evolution, 123f; explained using
    astragalus evolution, 134–35b, 136f; of
    hearing evolution, 141f; of skull and eye

position evolution, 152f; 165f, of
    protocetid whale relations, 165f; of
    bone and locomotor evolution, 186f; of
    tooth shape and jaw embryology
    evolution, 210f; of relationships of
    cetaceans to artiodactyls, 202f
clasper, 19
classification, 12–14, 14t, 215n18
clear-and-stain, 182f, 184f
Clementz, Mark, 123–24
cochlea, 139, 143
conglomerate, 39–41
continental crust, 69, 69f
continental drift, 70–71
continental shelf, 72
convergency, 131
core of Earth, 69, 69f
cortical layer, of bone, 151–53, 186f, 205.
    Synonym cortex
cow, 5, 42,82, 113, 128, 154, 159, 160,
    202f, Synonym cattle
cranial cavity, 25b, 115f, 139f
cranial nerve I, *see* olfactory nerve
cranial nerve II, *see* optic nerve
cranial nerve VII, *see* facial nerve
cranial nerve IX, *see* glossopharyngeal nerve
creationist, 4–5, 33, 47, 65, 116, 212
Crenatocetus, 163
Cretaceous period, 32f
cribriform plate, 115f
crocodiles, 3, 25, 59, 94, 108–09, 113, 120,
    142, 149, 154, 160, 214n4
crus breve, of incus, 7f, 8, 41, 137–39, 138f
crus longum, of incus, 7f, 8, 137–39, 138f
crust of Earth, 69–71, 69f
CT scan, 24b, 114, 115f, 168, 174

Dalanistes, 109, 111, 114
Darwin, Charles, 11–12, 19f, 33, 179
Davies, T.G.B., 38, 75, 76, 128
deciduous tooth, 23, 124. *Synonym* milk
    tooth
deer,7f, 8, 62f, 128, 134n, 167n, 200, 202f,
    204
Dehm, Richard, 38, 75–76, 128, 191–93,
    196–97
Delhi, India, 79–82
Delphinus, 13, 14, 27f, 62
delta value, in isotope geochemistry, 122,
    122f
dental formula, 21b, 22f, 111, 146, 166,
    203, 208, 215n21
desert pavement, 103
Dhedacetus, 165

heterodonty, 21, 210f
Heyning, John, 182, 209
hill station, 67
Himalayacetus, 58, 63, 83, 144
Himalayas, 2f, 32f, 67–75, 83–85, 220n1
hind limb: of *Ambulocetus,* 51, 64, 64f; of
    basilosaurids, 19, 29–30, 29f, 184–85;
    of bowhead whales, 29f, 30, 31f;
    development of, 177–87, 178f, 180f,
    182–84f, 186f; dolphin with, 173–76,
    176f, 185–87
Hindu Kush mountains, 2f, 71, 73
hippos, 5, 8, 25, 55f, 65, 113, 123f, 128,
    130, 134b, 135, 149, 151, 152f, 157,
    159, 171, 200–202, 202f. *Synonym*
    hippopotamus
Historia Animalium (Aristotle), 9
homodonty, 21, 209, 210f
hoof: of *Ambulocetus,* 64, 64f, 65; of
    Pakicetus, 153
hoofed mammal, 65, 128, 129, 153
horse, 15, 25b, 113, 116, 128
humerus, a bone of the forelimb, 104–05,
    127, 177, 182f, 186, 228n25
human, 12–14, 16, 21, 21b, 25b, 26, 28,
    28f, 29f, 32f, 53, 60–61b, 60f, 85, 89,
    94, 99, 121, 94, 134b, 142, 150, 158,
    167, 177, 181, 212, 218n8, 219n24
Hussain, Taseer, 68, 77
Hydrarchos harlandi, 17
Hydrargos sillimanii, 16–17, 16f
hydrofoil, 53
Hyemoschus, a mouse deer, 200
hyoid bone, 60f, 61b, 219n24
hypoglossal foramen, 90f

Ichthyolestes, 76, 144
Incisive foramen, 167b. *Synonym* anterior
    palatine foramen of fossil whales
incisor, 21, 21b, 22f, 59, 111, 147f, 167b,
    201, 204, 209, 210
incus, an ear ossicle, 7f, 8, 12, 33, 38, 137–39,
    138f, 141f, 142, 142f. *Synonym* anvil
India; 1992 expedition to, 79; 1996
    expedition to, 85; 2000 expedition to,
    99; 2005 expedition to, 191; map, 2f, 22f
India-Asia collision, 71–72
Indian antelope, 99
Indian Plate, 71, 72
Indocetus, 168
Indohyus, 199–205; astragalus, 133f; bone
    structure, 184f, 205; feeding and diet of,
    203–4; habitat and ecology of, 198, 205;
    incisors of, 201; involucrum of, 199,

199f; reconstruction of, 152f, 199–200,
    200–201f; location of, 200; osmoregula-
    tion, 123f; premolars of, 202; relation-
    ship of whales to, 199–203, 202f;
    skeleton of, 186f, 200f; skull of, 198–99,
    199f, 203–204; teeth of, 112f, 148f,
    201–2, 203–4; tooth wear in, 148f,
    150f, 202, 204; vision, 152f, 204;
    hearing of, 199–200, 204; walking and
    swimming by, 204–205
Indus Plain, 67, 74
Indus River, 2f, 71, 73
Indus Valley, 73–74
infraorbital canal, 90f
inner ear, 115f, 138f, 140f,
innominate, *see* pelvis
Institute of Cetacean Research, 188, 189
International Commission on Zoological
    Nomenclature, 12–14
International Whaling Commission (IWC),
    174, 189, 212
Intracoastal Waterway, 93
intelligence, 24b, 189
involucrum, 5f, 6, 33, 47, 114, 137, 138f,
    139, 142, 142f, 199–204, 199f
Islamabad, 2f, 3–4, 68, 79
isotope: carbon, 111, 132, 149, 165, 203,
    205; oxygen, 117–124, 117f, 123f;
    stable, 111, 117–24, 117f, 123f, 131,
    133, 146, 154, 205; stable vs. radio-
    genic, 117
isotope geochemistry, 117–24, 122f, 132,
    205
IWC (International Whaling Commission),
    174, 189, 212

Jacobson's organ, *see* vomeronasal organ
Japan, whaling in, 173–75, 187–89, 228n1
jaws: of ambulocetids, 45, 47, 58, 62, 62f;
    of basilosaurids, 62f; of *Ichthyolestes,*
    76; of *Indohyus,* 203–4; of *Pakicetus,*
    62f; of protocetids, 87, 163–66; of
    remingtonocetids, 86–87, 106, 111–13
jaw muscles, 23, 114. *Synonym* masticatory
    muscles, *see also* masseter, temporalis,
    and pterygoid muscle
jugal arch, 113. *Synonym* zygomatic arch
jugal bone, 90f. *Synonym* zygomatic bone
jugular foramen, 90f
Jurassic period, 32f, 36, 77, 144

K2 (mountain), 73
Kala Chitta Hills: *Ambulocetus* from,
    35–41, 58, 74; anthracobunid from,